Bayesian Inference - Recent Advantages

Edited by Niansheng Tang

Published in London, United Kingdom

Bayesian Inference - Recent Advantages
http://dx.doi.org/10.5772/intechopen.97942
Edited by Niansheng Tang

Contributors
Alessio De Francesco, Martin Boehm, Luisa Scaccia, Alessandro Cunsolo, Ali Mohammad-Djafari, Ahmed Saadoon, Frederick Bloetscher, Niansheng Tang, Ying Wu

First published in London, United Kingdom, 2022 by IntechOpen
IntechOpen is the global imprint of INTECHOPEN LIMITED, registered in England and Wales, registration number: 11086078, 5 Princes Gate Court, London, SW7 2QJ, United Kingdom

British Library Cataloguing-in-Publication Data
A catalogue record for this book is available from the British Library

Additional hard and PDF copies can be obtained from orders@intechopen.com

Bayesian Inference - Recent Advantages
Edited by Niansheng Tang
p. cm.
Print ISBN 978-1-80356-044-1
Online ISBN 978-1-80356-045-8
eBook (PDF) ISBN 978-1-80356-046-5

Meet the editor

Niansheng Tang is a professor of statistics and Dean of the School of Mathematics and Statistics, at Yunnan University, China. He was elected as a Yangtze River Scholars Distinguished Professor in 2013, as a member of the International Statistical Institute in 2016, as a member of the Board of Directors of the International Chinese Statistical Association (ICSA) in 2018, and as a Fellow of the Institute of Mathematical Statistics in 2021. He received the ICSA Outstanding Service Award in 2018 and the National Science Foundation Award for Distinguished Young Scholars of China in 2012. He is a member of the editorial board for *Statistics and Its Interface* and the *Journal of Systems Science and Complexity*. He is also a field editor for *Communications in Mathematics and Statistics*. His research interests include Bayesian statistics, missing data analysis, variable selection, high-dimensional data analysis, data science, and machine learning. He has published more than 200 research papers and authored seven books.

Contents

Preface

Due to the huge advances in computational and modeling techniques, Bayesian methods are becoming an increasingly important tool for analyzing various types of data such as continuous, discrete and mixed data, time-series data, longitudinal data, cross-sectional data, categorical data, survival data, missing data, high/ultrahigh-dimensional data, and latent variable data. They are also widely applied in various fields such as industry, agriculture, economics, engineering, medicine, biological ecology, social science, data science, machine learning, and AI for statistical inferences such as a parameter or non-parameter estimation, hypothesis testing, and prediction.

A variety of Bayesian inference theories, methodologies, and computational techniques have been developed due to the requirements for analyzing complicated data and models. These include structural, semi-structured, and unstructured data, as well as models without likelihoods or having a computing hard likelihood, times-series models, parametric or non/semi-parametric models, large-scale graphical or atmospheric models, network models, options pricing models, complex stochastic models, latent variable models, multilevel models, dynamic factor analysis models with/without time-varying parameters, high/ultrahigh-dimensional models, joint modeling of longitudinal and survival data, complex computer models, and causal inference models. Bayes factor computation, Bayesian variable selection, robust Bayesian inference, variational Bayesian inference, resampling, approximation of posterior distribution, approximate Bayesian computation, and debias methods are all of significance. But challenges remain with the development of AI and data mining requirements, such as how to balance computational time and statistical efficiency, design efficient Bayesian computational algorithms and robust sampling schemes for big/massive data, distributed data and streaming data, modeling and inference, while protecting privacy and guarding against malicious attacks.

This book, which features the work of five excellent researchers in theory, methods, models, algorithms and applications, has three sections and five chapters. Section I introduces the development of Bayesian inference, including theory, methods, algorithms and applications. Section II introduces Bayesian methods and includes two chapters. In Chapter 2, Professor Mohammad-Djafari Ali presents a Bayesian approach to solving inverse problems. In Chapter 3, Ph.D. candidate Ahmed Saadoon Mannaa investigates the prior data conflict by modeling the parameters in the prior distribution and comparing its standard deviation to that of the posterior distribution. A robust Bayesian method is presented that addresses the prior data conflict by using a set of prior distributions such as Weibull distribution and binomial distribution to identify the behavior of the estimators based on the estimation comparison of regular and robust Bayes methods via the integrated mean square error. The two chapters in Section III focus on the application of the hierarchical Bayesian method and optimized spectral acquisition in scattering experiments. In Chapter 4, Professor Bloetscher Frederick uses the hierarchical predictive Bayesian method to solve the challenge of "what to do when you have a complex question with numerous variables that are not well understood?". In Chapter 5, Alessio De Francesco, Luisa Scaccia, Marin Boehm and Alessandro Cuusolo suggest a Bayesian inference approach based on real-time analysis of experimental data and implemented as a series of steps in which the spectral measurement is adjourned by summing to its successive acquisition runs, and the spectral modeling is upgraded.

I was invited to edit this book after the publication of *Bayesian Analysis for Hidden Markov Factor Analysis Models*, which I co-wrote with Yemao Xia, Xiaoqian Zeng, and Niansheng Tang, and my two previously edited books, *Bayesian Inference on Complicated Data* and *Data Clustering,* published in 2020 and 2022, respectively. I am very grateful to Ms. Karla Skuliber for her kind invitation to edit this book and for providing me the chance to work with these authors. I sincerely hope this book will be of great interest to statisticians, data analysts, data scientists, social scientists, biologists, ecologists, and AI and machine learning researchers.

Niansheng Tang
Department of Statistics,
School of Mathematics and Statistics,
Yunnan University at Chenggong Campus,
Kunming, R. of China

Section 1

Introduction

Chapter 1

Introductory Chapter: Development of Bayesian Inference

Niansheng Tang and Ying Wu

1. Introduction

Bayesian inference derives from Bayesian theory, which depicts the probability of occurrence of an event given some prior information. Due to the huge advances in computational and modeling techniques, Bayesian inference has been increasingly become an important tool for data analysis in the Bayesian framework and has widely been applied to various fields, including social science, engineering, philosophy, medicine, sport, law and psychology, for parameter/nonparameter estimation, hypothesis test, and prediction. Various Bayesian methods including Markov chain Monte Carlo, objective Bayesian method, subjective Bayesian method, approximate Bayesian computation, and variational Bayesian methods have been developed to make Bayesian inference on various problems such as large-scale image classification and cluster analysis of microarray, and models including parametric, nonparametric, semiparametric models, and other complicated models such as joint models of survival data and longitudinal data, graphical models, computer models, neural network models, and spatial econometric models. In particular, in the big data era, various Bayesian fronts including theories, methods, and computational algorithms have been developed for accommodating the applications of AI and data science in recent years [1], for example, the prior learning, Bayes factor evaluation, Bayesian variable selection, robust Bayesian inference, variational Bayesian inference, resampling, approximation of posterior distribution, approximate Bayesian computation, and debias methods for high−/ultrahigh-dimensional data, multisource heterogeneous data, imbalanced data, missing data, and data stream. But there are some challenging problems, for example, how to balance the computational times and statistical efficiency, design efficient Bayesian computational algorithm and robust sampling schemes for big/massive data, distributed data and streaming data in the privacy protection and the defense of malicious attacks framework, and modeling so that they can adapt to the development of AI and the requirement of data mining, to be addressed and solved for Bayesian inference. In what follows, we will introduce the recent development and some topics of interest on Bayesian inference.

2. Bayesian estimation

For statistical models, Bayesian estimation is usually obtained from its posterior distribution based on Bayes theory. In general, Bayesian estimation includes Bayesian estimation of parameters and nonparametric functions. For parametric Bayesian estimation, we need first to specify the prior distribution of the parameter and then evaluate its posterior mean/median (i.e., Bayesian estimation of parameter) from its posterior distributions if the quadratic/mean-absolute loss function is used. For nonparametric Bayesian estimation, we need first approximate nonparametric function via some proper method such as B-splines and P-splines [2], i.e., parameterized approximation to nonparametric function, which leads to a parameterized model, and then employ Bayesian idea of parametric models to evaluate Bayesian estimates of parameter and nonparametric function. In what follows, we introduce how to evaluate Bayesian estimates of parameters or nonparametric functions in a relatively complicated model (e.g., random effects model/latent variable model).

In a latent variable model with missing response data, we assume the following form:

$$Y_i = X_i\beta + \Lambda\omega_i + \varepsilon_i, \quad \eta_i = \Pi\eta_i + \Gamma\xi_i + \epsilon_i, \quad i = 1,2,\dots,n, \tag{1}$$

where Y_i is a $p \times 1$ vector of manifest variables including continuous and categorical variables [3], X_i is a $p \times q$ matrix of covariates, ω_i is a $r \times 1$ vector of latent variables, ϵ_i and ε_i are the $p \times 1$ and $r_1 \times 1$ vectors of measurement errors, respectively, β is a $q \times 1$ vector of unknown parameters, Λ is a $p \times r$ factor loading matrix, $\omega_i = \left(\eta_i^T, \xi_i^T\right)^T$ in which η_i and ξ_i are the $r_1 \times 1$ and $r_2 \times 1$ sub-vectors of ω_i and $r_1 + r_2 = r$, respectively, Π and Γ are $r_1 \times r_1$ and $r_1 \times r_2$ matrices of unknown parameters. It is also assumed that $|I - \Pi| \neq 0$, ξ_i follows a normal distribution or an unknown distribution, and y_{ij}'s are subject to missingness, where I is a $r_1 \times r_1$ identity matrix, y_{ij} is the jth component of Y_i for $i = 1, \dots, n, j = 1, \dots, p$.

In general, a simple and standard assumption for the distributions of ε_i, ϵ_i, and ξ_i is to follow some parametric density family such as skew-normal/skew-t/skew-normal-cauchy/skew-symmetric-Laplace distribution [2] or exponential family distribution [3] or normal distribution [4], or unequal time autoregression series or their mixture. But this assumption may be unreasonable or too restrictive. To this end, some alternative methods have been developed to relax their parametric distribution assumption. For example, let ε_i or ξ_i follow an unknown distribution, which is approximated by a Dirichlet process prior [2], spiked Dirichlet process prior [5], truncated centered Dirichlet process prior [6], or a class of smooth densities that are approximated by the semiparametric approach of Gallant and Nychka [7], or unknown distribution with its quantiles specified leading to the well-known quantile regression models, or a Bayesian neural network approach to learning unknown distribution [8].

To introduce missing data, let δ_{ij} be an indicator of missing value y_{ij}, i.e., $\delta_{ij} = 1$ if y_{ij} is observed, and $\delta_{ij} = 0$ if y_{ij} is missing. In this case, we usually need to assume a missingness data mechanism, for example, missing at completely random or missing at random or missing not at random (also called nonignorable missing), and then specify a parametric or semiparmetric model for the considered missingness data mechanism. For example, see Lee and Tang [4] for the logistic regression model, Kim and Yu [9] and Tang, Zhao, and Zhu [10] for the exponential tilting model, and Wang and Tang [11] for the probit regression model together with latent variables. Also, one can train a missingness data mechanism model via a data-driven machine learning method [12], which is a completely new and not yet studied topic, including its implementation, algorithm, and theories. We are working on this new and promising topic, which may lead to a new research field.

To make Bayesian inference on the considered model (1), we need to specify a prior distribution for unknown parameters or coefficients in approximating unknown nonparametric functions. A standard assumption for unknown parameters is some proper parametric distribution family such as normal distribution, gamma distribution, inverse gamma distribution, inverse Gaussian distribution, Wishart distribution, and Beta distribution in which their hyperparameters are user prespecified. Their misspecification or improper application may lead to unreasonable even misleading parameter estimation. Bayesian inference based on these assumptions did not utilize historical data and limits its popularity in that the usage of historical data may improve the efficiency of parameter estimation. To address this issue, some relaxed priors are considered, for example, see power prior, g-prior, normalized power prior [13], calibrated power prior, dynamic power prior, and the power-expected-posterior prior and the scale transformed power prior [14]. For a high-dimensional sparse parametric model, we can assume a spike and slab prior for the parameter, which can be hierarchically expressed as a mixture of a normal distribution and an exponential distribution.

Let ϑ_1 be the set of unknown parameters associated with the distribution of ε_i, and ϑ_2 be the set of unknown parameters associated with the distributions of ε_i and ξ_i, ϑ_3 be the set of unknown parameters associated with the distribution of δ_i, and denote $\vartheta = \{\vartheta_1, \vartheta_2, \vartheta_3\}$. Let θ be a set of unknown parameters in $\{\beta, \Lambda, \Pi, \Gamma, \vartheta\}$. Denote $Y = \{Y_i, i = 1, \dots, n\}$, $Y_{\text{obs}} = \{Y_{i,\text{obs}}, i = 1, \dots, n\}$, $Y_{\text{mis}} = \{Y_{i,\text{mis}}, i = 1, \dots, n\}$, $X = \{X_i, i = 1, \dots, n\}$, $F = \{\omega_i, i = 1, \dots, n\}$, where $Y_{i,\text{mis}}$ and $Y_{i,\text{obs}}$ are sub-vectors of Y_i corresponding to missing and observed values, respectively. Let $D = \{Y_{\text{obs}}, X\}$ and $\delta = \{\delta_{ij}, i = 1, \dots, n, j = 1, \dots, p$. If the marginal posterior distribution of θ given the dataset D is

$$\pi((|\theta|D) = \int \pi(\theta, F, Y_{\text{mis}}, \delta|D)dFdY_{\text{mis}}d\delta = \frac{\int \pi(\theta)\pi(Y|\theta, F, X)\pi(F|\vartheta_2)\pi(\delta|Y, \vartheta_3 dFdY_{\text{mis}}d\delta}{\pi(Y_{\text{obs}}|X)}, \tag{2}$$

its posterior mean (i.e., Bayesian estimate) can be evaluated by

$$\hat{\theta} = E(\theta|D) = \int \theta\pi(\theta|D)d\theta = \frac{\int \theta\pi(\theta)\pi(Y|\theta, F, X)\pi(F|\vartheta_2)\pi(\delta|Y, \vartheta_3)dFdY_{\text{mis}}d\delta d\theta}{\pi(Y_{\text{obs}}|X)}, \tag{3}$$

where $\pi(\theta)$ is the prior distribution of θ, $\pi(Y|\theta, F, X)$ is the probability density function of Y given $(X, \beta, \Lambda, \vartheta_1)$, i.e., the likelihood function of θ associated with the considered latent variable model, $\pi(F|\vartheta_2) = \Pi_{i=1}^n \pi(\eta_i|\xi_i, \Pi, \Gamma, \vartheta_2)\pi(\xi_i|\vartheta_2)$ is the probability density of F, $\pi(Y_{\text{obs}}|X) = \int \pi(\theta)\pi(Y|\theta, F, X)\pi(F|\vartheta_2)\pi(\delta|Y, \vartheta_3)dFdY_{\text{mis}}d\delta d\theta$ is the marginal likelihood of Y_{obs} given X, and $\pi(\delta|Y, \vartheta_3)$ is the probability density of δ given (Y, ϑ_3).

From Eq. (3), it is easily seen that evaluating $\hat{\theta}$ is almost impossible due to high-dimensional integral involved. To address this issue, the well-known Markov chain Monte Carlo (MCMC) algorithm is employed to approximate $\hat{\theta}$ by sequentially drawing observations from the posterior distributions of components of θ and $(F, Y_{\text{mis}}, \delta)$ via the Gibbs sampler together with the Metropolis-Hastings algorithm. Denote $\pi(\theta|Y, F, X, \delta)$, $\pi(Y_{\text{mis}}|Y_{\text{obs}}, F, X, \theta, \delta)$, and $\pi(F|Y, \theta, X)$ as the conditional distributions of θ given (Y,F, X,δ), Y_{mis} given $(Y_{\text{obs}}, F, X, \delta)$, and F given (Y, θ, X), respectively. The Gibbs sampler is implemented as follows. At the tth iteration of the Gibbs sampler with the current observations $\left\{\theta^{(t)}, F^{(t)}, Y_{\text{mis}}^{(t)}, \delta^{(t)}\right\}$ of $\{\theta, F, Y_{\text{mis}}, \delta\}$, we sequentially draw i) $\beta^{(t+1)}$ from the conditional distribution $\pi(\beta|Y, F, X, \Lambda, \vartheta_1)$, ii) $\Lambda^{(t+1)}$ from the conditional distribution $\pi(\Lambda|Y, F, X, \beta, \vartheta_1)$, iii) $\Pi^{(t+1)}$ from the conditional distribution $\pi(\Pi|F, \Gamma, \vartheta_2)$, iv) $\Gamma^{(t+1)}$ from the conditional distribution $\pi(\Gamma|F, \Pi, \vartheta_2)$, v) $\vartheta_1^{(t+1)}$ from the conditional distribution $\pi(\vartheta_1|Y, X, \beta, \Lambda, F)$, vi) $\vartheta_2^{(t+1)}$ from the conditional distribution $\pi(\vartheta_2|F, \Pi, \Gamma)$, vii) $Y_{\text{mis}}^{(t+1)}$ from the conditional distribution $\pi(Y_{\text{mis}}|Y_{\text{obs}}, \delta, X, \beta, \Lambda, F, \vartheta_1)$, and viii) $F^{(t+1)}$ from the conditional distribution $\pi(F|Y, X, \theta)$. The aforementioned conditional distributions may be some familiar distributions from which observations can be directly drawn. But in some cases, these conditional distributions may be some unfamiliar and rather complicated distributions from which observations can be indirectly drawn. In this case, some alternative approaches, for example, the Metropolis-Hastings algorithm, rejection sampling, acceptance-rejection sampling, importance sampling, hybrid-jump-based sampling, and reversible jump sampling, can be employed to sample observations from these complicated distributions. The convergence of the above introduced Gibbs algorithm can be monitored by the estimated potential scale reduction (EPSR) values associated with the parameters [15], which are evaluated continuously as the iterations run. The Gibbs sampler converges if the EPSR values of unknown parameters are less than 1.2. Also, we can assess the convergence of the Gibbs sampler by plotting several parallel sequences of observations drawn from different starting values of unknown parameters against iterations.

Let $\{\theta^{(m)}, m = 1, \dots, M\}$, $\{F^{(m)}, m = 1, \dots, M\}$ and $\left\{Y_{\text{mis}}^{(m)}, m = 1, \dots, M\right\}$ be M observations sampled from their corresponding conditional distributions via the aforementioned Gibbs sampler after the Gibbs sampler algorithm converges, respectively. Bayesian estimates of θ, F, and Y_{mis} can be computed by

$$\hat{\theta} = \frac{1}{M}\sum_{m=1}^{M}\theta^{(m)}, \quad \hat{F} = \frac{1}{M}\sum_{m=1}^{M}\theta^{(m)}, \quad \hat{Y}_{\text{mis}} = \frac{1}{M}\sum_{m=1}^{M}Y_{\text{mis}}^{(m)}, \tag{4}$$

respectively. Their corresponding standard deviations can be computed with their corresponding sample covariance matrices of the observations. The details can refer to the literature [7]. The above argument on Bayesian inference is a classical method. However, for a high-dimensional parametric or nonparametric model, one needs some new approaches to solve the computing time and efficiency and stability of algorithm problem. In fact, when the dimension of covariate matrix is large and the sample size is relatively small, i.e., the well-known "large p and small n" problem, the Gibbs sampler is computationally expensive and has poor stability.

To solve this issue for a high-dimensional regression model, there are some novel approaches developed for parameter/nonparametric function estimation in the Bayesian framework, for example, see Bayesian Lasso, Bayesian adaptive Lasso, Bayesian elastic net, and Bayesian $L_{1/2}$. These approaches can be utilized to estimate model parameters or nonparametric functions and are simultaneously used to select variables in a high-dimensional regression model, which have received considerable attention and extended to various models such as generalized linear models and linear mixed models. In particular, to reduce the computational cost, various variational Bayesian methods have been developed for various models in recent years. For example, see linear mixed models [16] and reference therein. However, there are a lot of unsolved problems. For example, for a complicated model, how to find the optimal variational densities for approximating complicated posterior distributions? How to extend/break the assumption of mean field that is a basic assumption in variational analysis? How to utilize other divergence criteria rather than Kullback–Leibler divergence to develop variational Bayesian theories?

3. Model comparison

Model comparison is widely used to select a plausible model to fit a given dataset among all the considered candidate models. Various methods have been developed to make model comparisons for many models such as linear/nonlinear regression models, structural equation models, multilevel models, machine learning models, and pattern recognition model in the Bayesian framework over the past years.

To select a better model among all the candidate models, we can adopt the well-known best subset selection methods such as Akaike information criterion (AIC), Bayesian information criterion (BIC), deviance information criterion (DIC), generalized information criterion (GIC), minimum description length (MDL), Hannan-Quinn information criterion (HIC), and log scoring criterion (also called the conditional predictive ordinate, i.e., CPO), which trade off a measure of model plausibility and a measure of model complexity. Also, the Bayes factor [17] has been developed to conduct Bayesian model comparison and is widely utilized to investigate the strength of the evidence in favor of one model among two candidate models. The Bayes factor for two competing models H_0 and H_1 is defined as follows:

$$\mathrm{B}_{10} = \frac{\pi(Y_{\mathrm{obs}}, \delta|H_1)}{\pi(Y_{\mathrm{obs}}, \delta|H_0)},$$

where $\pi(Y_{\mathrm{obs}}, \delta|H_k) = \int \pi(Y_{\mathrm{obs}}, \delta|\theta_k)\pi(\theta_k)d\theta_k$ is the marginal density of H_k with parameter vectors θ_k, and $\pi(\theta_k)$ is the prior density of θ_k associated with model H_k for $k = 0,1$. In general, if the Bayes factor $\mathrm{B}_{10} > 1$, the model H_1 is more plausible by the observed data than the model H_0, which leads to the following model comparison rule: B_{10}'s value lying in the intervals (3,10), (10,30), (30,100), and (100,∞) yields moderate, strong, very strong, and extreme evidence in favor of model H_1, respectively. It is rather difficult to compute $\pi(Y_{\mathrm{obs}}, \delta|\theta_k)$ due to the intractable high-dimensional integral involved, thus computing the Bayes factor B_{10} is challenging. Many methods have been proposed to compute marginal likelihoods $\pi(Y_{\mathrm{obs}}, \delta|H_k)$ or Bayes factors [3]. For example, see importance sampling, path sampling, bridge sampling, Harmonic mean method, random weight importance sampling, sequential Monte Carlo method, and pareto-smoothed importance sampling leave-one-out cross-validation.

One serious defect of the Bayes factor for model comparison is that it is well defined for improper priors of θ_k's and is sensitive to the selection of the hyperparameters in the priors. According to our experience, different priors together with different sampling methods lead to different values of the Bayes factor, i.e., different model comparison results. To this end, some modifications of the Bayes factor have been proposed, for instance, the partial Bayes factor, the intrinsic Bayes factor, and the fractional Bayes factor, which are subject to more or less arbitrary selection of training samples, weights for averaging training samples, and fractions, respectively. Also, some robust methods were developed to compute the sensitivity of the marginal likelihoods via the simulation-based methods, called the automated prior robustness method. Recently, some novel methods were proposed to deal with improper priors in computing the Bayes factor. For example, see machine learning method, i.e., first using a part of the dataset studied to train the Bayes factor/transform the improper prior into a proper prior and then utilizing the remainder of the dataset for model comparison, which provides a new idea for computing the Bayes factor. The robustness of model comparison is a challenging topic, which is worth further studying.

Author details

Niansheng Tang* and Ying Wu
School of Mathematics and Statistics, Yunnan University Chenggong, Kunming, China

*Address all correspondence to: nstang@ynu.edu.cn

References

[1] Tang N, Liu C, Shi JQ, Huang Y. Editorial: Bayesian inference and Ai. Frontiers in Big Data. 2022;**5**:1-2

[2] Tang AM, Tang NS. Semiparametric Bayesian inference on skew-normal joint modeling of multivariate longitudinal and survival data. Statistics in Medicine. 2015;**34**:824-843

[3] Lee SY, Tang NS. Bayesian analysis of structural equation models with mixed exponential family and ordered categorical data. British Journal of Mathematical and Statistical Psychology. 2006;**59**:151-172

[4] Lee SY, Tang NS. Analysis of nonlinear structural equation models with nonignorable missing covariates and ordered categorical data. Statistica Sinica. 2006;**16**:1117-1141

[5] Kim S, Dahl DB, Vannucci M. Spiked Dirichlet process prior for Bayesian multiple hypothesis testing in random effects models. Bayesian Analysis. 2009; **4**:707-732

[6] Tang N, Wu Y, Chen D. Semiparametric Bayesian analysis of transformation linear mixed models. Journal of Multivariate Analysis. 2018; **166**:225-240

[7] Gallant AR, Nychka DW. Semiparametric maximum likelihood estimation. Econometrica. 1987;**55**: 363-390

[8] Wright WA. Bayesian approach to neural-network modeling with input uncertainty. IEEE Transactions on Neural Network. 1999;**10**:1261-1270

[9] Kim JK, Yu CL. A semeiparametric estimation of mean functionals with nonignorable missing data. Journal of the American Statistical Association. 2011; **2011**(106):157-165

[10] Tang NS, Zhao PY, Zhu H. Empirical likelihood for estimating equations with nonignorable missing data. Statistica Sinica. 2014;**24**:723-747

[11] Wang ZQ, Tang NS. Bayesian quantile regression with mixed discrete and nonignorable missing covariates. Bayesian Analysis. 2020;**15**:579-604

[12] Liu M, Zhang Y, Zhou D. Double/ debiased machine learning for logistic partially linear model. The Econometrics Journal. 2021;**24**:559-588

[13] Ibrahim JG, Chen MH, Sinha D. On optimality properties of the power prior. Journal of the American Statistical Association. 2003;**98**:204-213

[14] Nifong B, Psioda MA, Ibrahim JG. The scale transformed power prior for use with historical data from a different outcome model. DOI: 10.48550/ arXiv.2105.05157

[15] Gelman A. Inference and monitoring convergence. In: Gilks WR, Richardson S, Spiegelhalter DJ, editors. Markov Chain Monte Carlo in Practice. London: Chapman and Hall; 1996. pp. 131-143

[16] Yi JY, Tang N. Variational Bayesian inference in high-dimensional linear mixed models. Mathematics. 2022;**10**: 463

[17] Kass RE, Raftery AE. Bayes factors. Journal of the American Statistical Association. 1995;**90**:773-795

Section 2

Bayesian Inference Methods

Chapter 2

Bayesian Inference for Inverse Problems

Ali Mohammad-Djafari

Abstract

Inverse problems arise everywhere we have indirect measurement. Regularization and Bayesian inference methods are two main approaches to handle inverse problems. Bayesian inference approach is more general and has much more tools for developing efficient methods for difficult problems. In this chapter, first, an overview of the Bayesian parameter estimation is presented, then we see the extension for inverse problems. The main difficulty is the great dimension of unknown quantity and the appropriate choice of the prior law. The second main difficulty is the computational aspects. Different approximate Bayesian computations and in particular the variational Bayesian approximation (VBA) methods are explained in details.

Keywords: inverse problems, hidden variable, hierarchical models, approximate Bayesian computation, variational Bayesian approximation

1. Introduction

Inverse problems arise in many scientific and engineering applications. In fact, almost always we want to infer on quantities, variables, distributions and functions which are not directly observable. Inferring on a hidden variable f via the observation of another variable g is the main objective in many scientific area [1–3].

Classically, the Bayesian methods have been developed for direct observation of a quantity, its parametric modeling and the estimation of the parameters of the model. Description and development of the Bayesian inference for the case of inverse problems is the main objective of this chapter. The chapter is organized as follows: First a few inverse problems are mentioned, mainly in two categories: those described by Ordinary Differential Equations (ODE) and Partial Differential Equations (PDE)'s and those described with integral equations. Then, we will see that two main problems arise: parameter estimation and inversion. First the Bayesian parameter estimation is described and then the Bayesian inversion.

2. Inverse problems

To see easily the two categories of inverse problems, first a very simple example is given. Consider the electrical circuit of **Figure 1**.

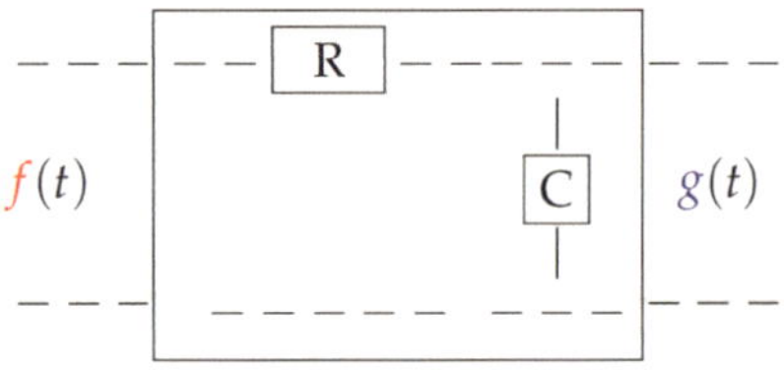

$$\frac{\partial g(t)}{\partial t} = \frac{1}{C} i(t), \quad i(t) = \frac{(f(t) - g(t))}{R}, \quad \frac{\partial g(t)}{\partial t} = \frac{1}{RC}(f(t) - g(t))$$

$$g(t) + RC\frac{\partial g(t)}{\partial t} = f(t) \longrightarrow g(\omega) + RCj\omega g(\omega) = F(\omega)$$

$$H(\omega) = \frac{g(\omega)}{F(\omega)} = \frac{1}{1 + jRC\omega}$$

$$RC = \theta \longrightarrow H(\omega) = \frac{1}{1 + j\theta\omega} \rightarrow h(t) = \exp\left[-t/\theta\right] \rightarrow f(t) = h(t) * f(t)$$

Figure 1.
A simple electrical circuit example to show two different expressions of inverse problems modeling: Ordinary differential equation (ODE) or Integral equation (IE).

Using the notations used on the figure, we can easily obtain the following ODE:

$$g(t) + \boldsymbol{\theta}\frac{\partial g(t)}{\partial t} = f(t) \tag{1}$$

Then, using the Fourier transform (FT), we obtain easily the following integral equation:

$$f(t) = \int f(\tau) h(t - \tau)\, \mathrm{d}\tau \tag{2}$$

These two simple equations describe the same linear inverse problem, where we can distinguish the following mathematical problems:

- Forward problem: Given the parameter $\boldsymbol{\theta}$ of the system and the input, $f(t)$ predict the output $g(t)$.
- Parameter estimation: Given the input $f(t)$ and the output $g(t)$, estimate the parameter $\boldsymbol{\theta}$.
- System identification: Given the input $f(t)$ and the output $g(t)$, estimate the impulse response (IR) of the system $h(t)$.
- Inverse problems:
  - Simple: Given the characteristics of the system (either the parameter $\boldsymbol{\theta}$ or equivalently the impulse response $h(t)$) and the output $g(t)$ estimate the input $f(t)$;
  - Blind: Given the output $g(t)$ estimate both the system, parameter $\boldsymbol{\theta}$ or the impulse response $h(t)$, and input $f(t)$.

For general vocabulary and examples, see [2, 4, 5].

2.1 Examples of linear inverse problems

Here, a few examples of classical inverse problems are listed.

2.1.1 Deconvolution

When the forward problem is a convolution operation:

$$g(t) = \int f(\tau)h(t-\tau)\,\mathrm{d}\tau, \tag{3}$$

the inverse problem is called *Deconvolution* (**Figure 2**). **Figure 3** shows an example of deconvolution problem which arise in radio astronomy.

2.1.2 Image restoration

In many imaging systems, such as visual cameras, microscopes, telescopes or Infra Red cameras, due to some limitations such as limited aperture or limited resolution, the forward problem can be approximated by a 2D convolution equation:

$$g(x,y) = \iint f(x',y')h(x-x',y-y')\,\mathrm{d}x\,\mathrm{d}y. \tag{4}$$

The corresponding inverse problem is called image deconvolution or more often *image restoration*. The example given in **Figure 4**, is the case of satellite imaging [6, 7].

Figure 2.
Signal deconvolution problem.

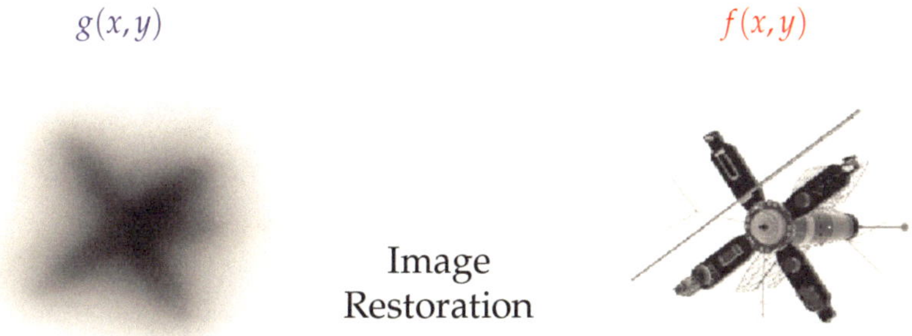

Figure 3.
Image deconvolution or restoration inverse problem in sattelite imaging.

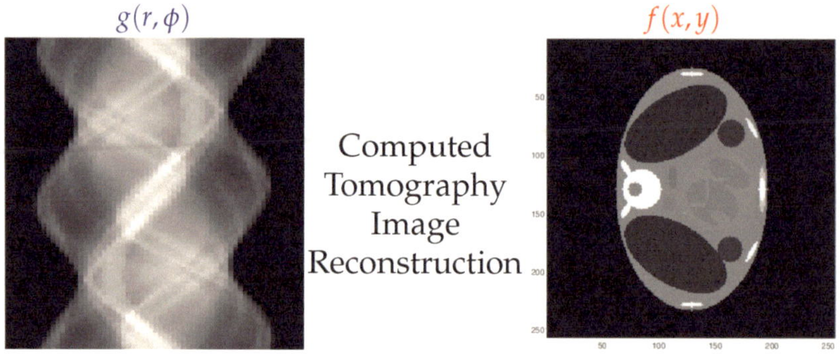

Figure 4.
Image reconstruction in CT. On the left, the projections $g(r, \phi)$ and on the right the object $f(x, y)$.

2.1.3 Image reconstruction in X ray computed tomography (CT)

In X-ray CT, assuming parallel geometry, where a ray is characterized by its angle ϕ and its distance r from the center of the object $f(x, y)$ the relation between the data $g(r, \phi)$, called projections at angle ϕ and the function $f(x, y)$, called object, is given by the Radon transform:

$$g(r, \phi) = \iint f(x, y) \delta(r - x \cos \phi - y \sin \phi) \, dx \, dy. \tag{5}$$

The inverse problem here is called *Image reconstruction*. A simulated example is shown in **Figure 4**.

2.1.4 Fourier synthesis

In many imaging systems, when using FT, it is possible to model the inverse problem with the following forward FT relation:

$$g(u, v) = \iint f(x, y) \exp \left[-j(ux + vy)\right] \, dx \, dy, \tag{6}$$

where the data, after an appropriate FT, can fill partially the Fourier domain $g(u, v)$ of the unknown interested function $f(x, y)$ [8, 9]. **Figure 5** shows the case of X-ray CT.

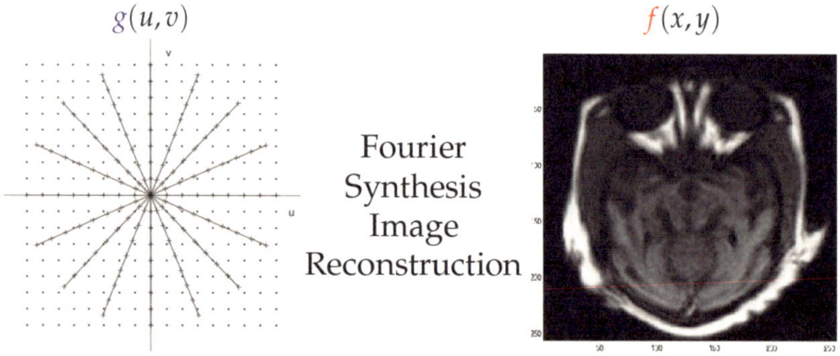

Figure 5.
Fourier synthesis (FS) inverse problems arising in many imaging systems. Here is illustrated the FS problem in X-ray CT.

2.1.5 General linear inverse problems

All the examples of the linear inverse problems listed above, can be summarized in the following general form:

$$g(\boldsymbol{s}) = \int f(\boldsymbol{r})\, h(\boldsymbol{s}, \boldsymbol{r})\, \mathrm{d}\boldsymbol{r} \tag{7}$$

where $\boldsymbol{s}$ can be either t, (x, y), (r, ϕ) or (u, v) and $\boldsymbol{r}$, respectively τ, (x', y'), (x, y) and (x, y).

3. Bayesian parameter estimation

To introduce, in a very simple way, the Bayes rule for parameter estimation, we consider the case where we have a set of data: $\boldsymbol{g} = \{g_1, \cdots, g_n\}$ where we assign them a probability law $p(g_i|\boldsymbol{\theta})$ with a set of unknown parameters $\boldsymbol{\theta}$. The question now is how to infer $\boldsymbol{\theta}$ from those data. We can immediately use the Bayes rule:

$$p(\boldsymbol{\theta}|\boldsymbol{g}) = \frac{p(\boldsymbol{g}|\boldsymbol{\theta})p(\boldsymbol{\theta})}{p(\boldsymbol{g})} \propto l(\boldsymbol{\theta})p(\boldsymbol{\theta}) \tag{8}$$

where:

- $l(\boldsymbol{\theta}) \triangleq p(g|\boldsymbol{\theta}) = \Pi_i p(g_i|\boldsymbol{\theta})$ is called the *likelihood*, representing the uncertainty in the data knowing the parameters;
- $p(\boldsymbol{\theta})$ is called the *prior or* a priori, a probability law assigned to the parameters to represent the prior knowledge (to the observation data) we may have on those parameters;
- the denominator $p(\boldsymbol{g})$

$$p(\boldsymbol{g}) = \int p(g|\boldsymbol{\theta})p(\boldsymbol{\theta})\, \mathrm{d}\boldsymbol{\theta} \tag{9}$$

is called the *evidence*.

So, the process of using the Bayes rule for parameter estimation can be summarized as follows:

- Write the expression of the likelihood $p(\boldsymbol{g}|\boldsymbol{\theta})$
- Assign the prior $p(\boldsymbol{\theta})$ to translate all we know about $\boldsymbol{\theta}$ before observing the data $\boldsymbol{g}$
- Apply the Bayes rule to obtain the expression of the posterior law $p(\boldsymbol{\theta}|\boldsymbol{g})$
- Use the posterior $p(\boldsymbol{\theta}|\boldsymbol{g})$ to do any inference on $\boldsymbol{\theta}$. For example:
    - Compute its expected value, called Expected A Posteriori (EAP) or Posterior Mean (PM):

$$\hat{\boldsymbol{\theta}}_{PM} = \int \boldsymbol{\theta} p(\boldsymbol{\theta}|\mathbf{g})\, \mathrm{d}\boldsymbol{\theta} \tag{10}$$

- Compute the value of $\boldsymbol{\theta}$ for which the $p(\boldsymbol{\theta}|\mathbf{g})$ is maximum; Maximum A Posteriori (MAP):

$$\hat{\boldsymbol{\theta}}_{MAP} = \arg\max_{\boldsymbol{\theta}} \{p(\boldsymbol{\theta}|\mathbf{g})\} \tag{11}$$

- Sampling and exploring [Monte Carlo methods]

$$\boldsymbol{\theta} \sim p(\boldsymbol{\theta}|\mathbf{g})$$

which gives the possibility to obtain any statistical information we want to know about $\boldsymbol{\theta}$. For example, if we generate N samples $\{\boldsymbol{\theta}_1, \cdots, \boldsymbol{\theta}_N\}$, for large enough N, we have:

$$\mathrm{E}\{\theta\} \simeq \frac{1}{N} \sum_{n=1}^{N} \theta_n. \tag{12}$$

3.1 One parameter case

When θ is a scalar quantity, then, we can also do the following computations:

- Compute the value of θ_{Med} such that:

$$P(\theta > \theta_{Med}) = P(\theta < \theta_{Med}) \tag{13}$$

which is called the *median value*. Its computation needs integration:

$$\int_{-\infty}^{\theta_{Med}} p(\theta|\mathrm{g})\, \mathrm{d}\theta = \int_{\theta_{Med}}^{\infty} p(\theta|\mathbf{g})\, \mathrm{d}\theta \tag{14}$$

- Compute the value θ_α, called α quantile, for which

$$P(\theta > \theta_\alpha) = 1 - P(\theta < \theta_\alpha) = \int_{\hat{\theta}_\alpha}^{\infty} p(\theta|\mathbf{g})\, \mathrm{d}\theta = 1 - \alpha \tag{15}$$

- Region of high probabilities: [needs integration methods]

$$[\hat{\theta}_1, \hat{\theta}_2] : \int_{\hat{\theta}_1}^{\hat{\theta}_2} p(\theta|\mathbf{g})\, \mathrm{d}\theta = 1 - \alpha$$

Bayes rule and Bayesian estimation can be illustrated as follows:

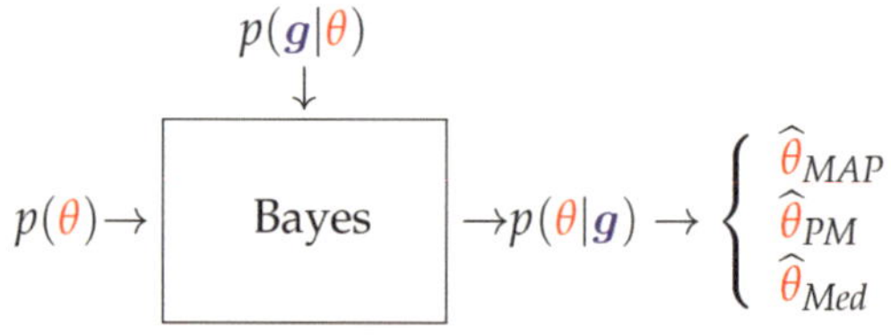

Two main points are of great importance:

- How to assign the prior $p(\boldsymbol{\theta})$ in the second step; and
- How to do the computations in the last step.

This last problem becomes more serious with multi parameter case.

3.2 Multi-parameter case

If we have more than one parameter, then $\boldsymbol{\theta} = [\theta_1, \cdots, \theta_n]'$. The Bayes rule still holds:

$$p(\boldsymbol{\theta}|g) = \frac{p(\mathbf{g}|\boldsymbol{\theta})p(\boldsymbol{\theta}|\mathbf{g})}{p(\mathbf{g})} \tag{16}$$

Now, again, we can compute:

- The Expected A Posteriori (EAP):

$$\hat{\boldsymbol{\theta}}_{PM} = \int \boldsymbol{\theta} p(\boldsymbol{\theta}|\mathbf{g})\, \mathrm{d}\boldsymbol{\theta}, \tag{17}$$

but this needs efficient integration methods.

- The Maximum A Posteriori (MAP):

$$\hat{\boldsymbol{\theta}}_{MAP} = \arg\max_{\boldsymbol{\theta}} \{p(\boldsymbol{\theta}|\mathbf{g})\} \tag{18}$$

but this needs efficient optimization methods.

- Sampling and exploring [Monte Carlo methods]

$$\boldsymbol{\theta} \sim p(\boldsymbol{\theta}|\mathbf{g})$$

but this needs efficient sampling methods.

- We may also try to localize the region of the highest probability:

$$P(\boldsymbol{\theta} \in \Theta) = \int_{\Theta} p(\boldsymbol{\theta}|g)\, \mathrm{d}\boldsymbol{\theta} = 1 - \alpha \tag{19}$$

for a given small α, but this problem may not have a unique solution.

4. Bayesian inference for inverse problems

As described before, in inverse problems, the unknown f is a function (of time, space, wavelength, ...) and the observable quantity g is also another function which is related to f via an operator $g = \mathcal{H}(f) + \epsilon$. When discretized, they can be represented by the great dimensional vectors $\boldsymbol{f}, \mathbf{g}$ and $\mathbf{g} = \boldsymbol{H}(\boldsymbol{f}) + \boldsymbol{\epsilon}$, where ϵ represents all the errors (measurement, model and discretization). When the operator is a linear one, we have:

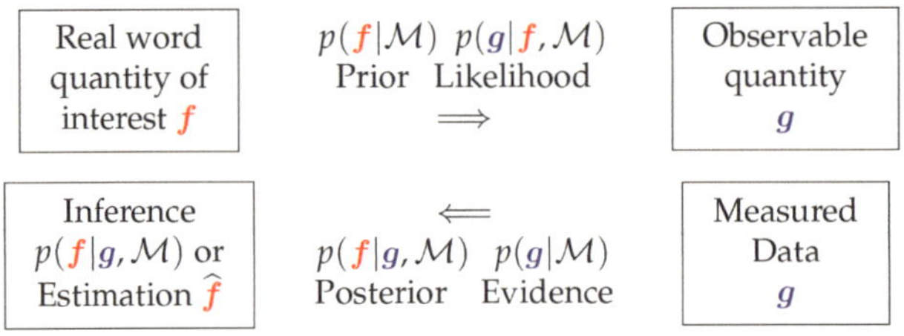

$$p(f|g,\mathcal{M}) = \frac{p(g|f,\mathcal{M})\; p(f|\mathcal{M})}{p(g|\mathcal{M})} \quad \text{with } p(g|\mathcal{M}) = p(g|f,\mathcal{M})\; p(f|\mathcal{M})\,\mathrm{d}f$$

Figure 6.
Illustration of the Bayesian inference for inverse problems.

$$\boldsymbol{g} = \boldsymbol{H}\boldsymbol{f} + \boldsymbol{\epsilon}, \tag{20}$$

where $\boldsymbol{f}$ is a vector of length n, $\boldsymbol{H}$ the known forward model matrix of size $m \times n$, and $\boldsymbol{g}$ and $\boldsymbol{\epsilon}$ two vectors of size m.

The Bayes rule for this case is written as:

$$p(\boldsymbol{f}|\boldsymbol{g},\mathcal{M}) = \frac{p(\boldsymbol{g}|\boldsymbol{f},\mathcal{M})p(\boldsymbol{f}|,\mathcal{M})}{p(\boldsymbol{g},\mathcal{M})} \tag{21}$$

where we introduce $\mathcal{M}$ to represent the model, $p(\boldsymbol{g}|\boldsymbol{f},\mathcal{M})$, called commonly the *likelihood*, is obtained using the forward model (20) and the assigned probability law of the noise $p(\boldsymbol{\epsilon}), p(\boldsymbol{f}|,\mathcal{M})$ is the assigned prior model and $p(\boldsymbol{f}|\boldsymbol{g},\mathcal{M})$ the posterior probability law. **Figure 6** shows in a schematic way the main ingredients of the Bayesian inference for inverse problems.

This even very simple linear model has been used in many areas: linear inverse problems, compressed sensing, curve fitting and linear regression, machine learning, etc.

In inverse problems such as deconvolution, image restoration, $\boldsymbol{f}$ represent the input or original image, $\boldsymbol{g}$ represents the blurred and noisy image and $\boldsymbol{H}$ is the convolution operator matrix. In image reconstruction in Computed Tomography (CT), $\boldsymbol{f}$ represents the distribution of some internal property of an object, for example the density of the material and $\boldsymbol{g}$ represents the radiography data and $\boldsymbol{H}$ is the radiographic projection operator (discretized Radon transform operator).

In Compress Sensing, $\boldsymbol{g}$ is the compressed data, $\boldsymbol{f}$ is the uncompressed image and $\boldsymbol{H}$ the compressing matrix. In machine learning, $\boldsymbol{g}$ are the data, $\boldsymbol{H}$ is a dictionary and $\boldsymbol{f}$ represents the sparse coefficients of the projections of the data on that dictionary.

5. Hyperparameter estimation

When applying the Bayes rule, the main terms which are the likelihood and prior depend on parameters, which cannot be fixed in practical situation. We may thus want to estimate them from the data. In the Bayesian approach, this can be done easily:

$$p(\boldsymbol{f},\boldsymbol{\theta}_1,\boldsymbol{\theta}_2|\boldsymbol{g}) = \frac{p(\boldsymbol{g}|\boldsymbol{f},\boldsymbol{\theta}_1)p(\boldsymbol{f},\boldsymbol{\theta}_2)p(\boldsymbol{\theta}_1)p(\boldsymbol{\theta}_2)}{p(\boldsymbol{g})} \tag{22}$$

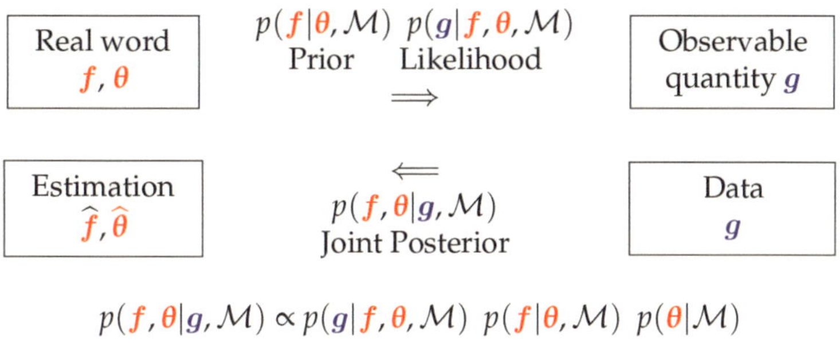

Figure 7.
Illustration of the Bayesian approach for inverse problems with unknown hyperparameters.

where $p(\boldsymbol{\theta}_1)$ and $p(\boldsymbol{\theta}_2)$ are the prior probability laws assigned to $\boldsymbol{\theta}_1$ and $\boldsymbol{\theta}_2$ and often $p(\boldsymbol{\theta}) = p(\boldsymbol{\theta}_1)p(\boldsymbol{\theta}_2)$. We can then write more succinctly:

$$p(\boldsymbol{f}, \boldsymbol{\theta}|\boldsymbol{g}, \boldsymbol{\theta}_0) = \frac{p(\boldsymbol{g}|\boldsymbol{f}, \boldsymbol{\theta}_1)p(\boldsymbol{f}, \boldsymbol{\theta}_2)p(\boldsymbol{\theta})}{p(\boldsymbol{g})} \tag{23}$$

The scheme of this situation is illustrated in **Figure 7**.

From here, we have different directions for doing estimation:

5.1 Joint maximum a posteriori (JMAP)

Rewriting the expression of the joint posterior law:

$$p(\boldsymbol{f}, \boldsymbol{\theta}|\boldsymbol{g}) = \frac{p(\boldsymbol{g}|\boldsymbol{f}, \boldsymbol{\theta}_1)p(\boldsymbol{f}|\boldsymbol{\theta}_2)p(\boldsymbol{\theta})}{p(\boldsymbol{g})} \propto p(\boldsymbol{g}|\boldsymbol{f}, \boldsymbol{\theta}_1)p(\boldsymbol{f}, \boldsymbol{\theta}_2)p(\boldsymbol{\theta}) \tag{24}$$

where $\propto$ means equal up to a constant factor which is $1/p(\boldsymbol{g})$. In this case, we can try to optimize it with respect to its two arguments:

$$\left(\hat{\boldsymbol{f}}, \hat{\boldsymbol{\theta}}\right) = \arg\max_{(\boldsymbol{f}, \boldsymbol{\theta})} \left\{p\left(\boldsymbol{f}, \hat{\boldsymbol{\theta}}|\boldsymbol{g}\right)\right\} \tag{25}$$

This can be done, for example, by alternate optimization:

$$\begin{cases} \hat{\boldsymbol{f}}^{(k+1)} = \arg\max_{\boldsymbol{f}} \left\{p\left(\boldsymbol{f}, \hat{\boldsymbol{\theta}}^{(k)}|\boldsymbol{g}\right)\right\} \\ \hat{\boldsymbol{\theta}}^{(k+1)} = \arg\max_{\boldsymbol{\theta}} \left\{p\left(\boldsymbol{f}^{(k)}, \hat{\boldsymbol{\theta}}|\boldsymbol{g}\right)\right\} \end{cases} \tag{26}$$

When the optimization algorithm is successful, we have the optimal values of $\hat{\boldsymbol{f}}$ and $\hat{\boldsymbol{\theta}}$. This method can be summarized as follows:

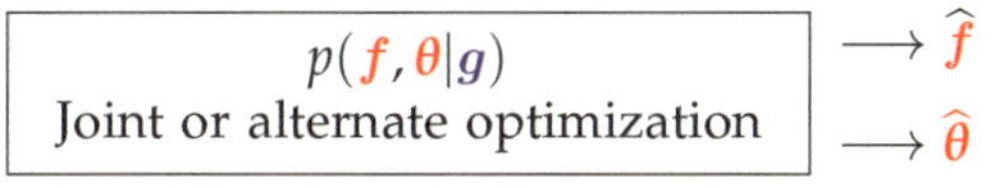

5.2 Marginalization over θ

The main idea here is to consider $\boldsymbol{\theta}$ as a nuisance parameter. Thus, integrating it out, we get

$$p(\boldsymbol{f}|\mathbf{g}) = \int p(\boldsymbol{f},\boldsymbol{\theta}|\mathbf{g})\, \mathrm{d}\boldsymbol{\theta} \tag{27}$$

which can be used to infer on $\boldsymbol{f}$. Also, if we still want to get estimates of $\boldsymbol{\theta}$, we can first obtain an estimate $\hat{\boldsymbol{f}}$ for $\boldsymbol{f}$ and then, if needed, to use it as it is illustrated in the following scheme:

$$\boxed{p(f,\theta|g)} \longrightarrow \boxed{p(f|g)} \longrightarrow \widehat{f} \longrightarrow \boxed{p(\theta|\widehat{f},g)} \longrightarrow \widehat{\theta}$$

Joint Posterior Marginalize over θ

5.3 Marginalization over f

The main idea here is first find a good estimate for the parameters $\boldsymbol{\theta}$ and then use it for the inference on $\boldsymbol{f}$. So, first obtain:

$$p(\boldsymbol{\theta}|\mathbf{g}) = \int p(\boldsymbol{f},\boldsymbol{\theta}|\mathbf{g})\, \mathrm{d}\boldsymbol{f} \tag{28}$$

which can be used to first estimate $\boldsymbol{\theta}$ and then use it. For example, the method which is related to the *Second type Maximum likelihood*, first estimate $\hat{\boldsymbol{\theta}}$ by

$$\hat{\boldsymbol{\theta}} = \arg\max_{\boldsymbol{\theta}} \left\{p\left(\hat{\boldsymbol{\theta}}|\mathbf{g}\right)\right\} \tag{29}$$

and then use it with $p\left(\boldsymbol{f}|\hat{\boldsymbol{\theta}},\mathbf{g}\right)$ to infer on $\hat{\boldsymbol{f}}$. For a flat prior model, $p(\boldsymbol{\theta}|g) \propto p(g|\boldsymbol{\theta})$ which is called the *likelihood* and the estimator

$$\hat{\boldsymbol{\theta}} = \arg\max_{\boldsymbol{\theta}} \left\{p\left(\hat{\boldsymbol{\theta}}|g\right)\right\} = \arg\max_{\boldsymbol{\theta}} \left\{p\left(g|\hat{\boldsymbol{\theta}}\right)\right\} \tag{30}$$

is called *Maximum Likelihood (ML)* and the whole approach is called *ML of second type*. This method can be summarized as follows:

$$\boxed{p(f,\theta|g)} \longrightarrow \boxed{p(\theta|g)} \longrightarrow \widehat{\theta} \longrightarrow \boxed{p(f|\widehat{\theta},g)} \longrightarrow \widehat{f}$$

Joint Posterior Marginalize over f

The main difficulty in this approach is that, rarely we can have an analytical expression for the first marginalization. To overcome this difficulty, many algorithms have been proposed to compute $\boldsymbol{f}$. One of them is called Expectation- Maximization (EM) and its generalization (GEM). The main idea of these algorithms are summarized in the following subsections:

5.3.1 EM and GEM algorithms

To summarize these methods, we use the vocabulary of the main authors of EM method, where $\boldsymbol{f}$ is considered as *hidden variable*, $\boldsymbol{g}$ as *incomplete data*, $(\boldsymbol{g},\boldsymbol{f})$ as *complete data*, $\ln p(\boldsymbol{g}|\boldsymbol{\theta})$ *incomplete data log-likelihood* and $\ln p(\boldsymbol{g},\boldsymbol{f}|\boldsymbol{\theta})$ as *complete data log-likelihood*. Then, the following iterative algorithms describe the EM and GEM algorithms.:

- EM Iterative algorithm:

$$\begin{cases} \text{E-step}: & Q\left(\boldsymbol{\theta},\hat{\boldsymbol{\theta}}^{(k)}\right) = \mathrm{E}_{p\left(\boldsymbol{f}|\boldsymbol{g},\hat{\boldsymbol{\theta}}^{(k)}\right)}\{\ln p(\boldsymbol{g},\boldsymbol{f}|\boldsymbol{\theta})\} \\ \text{M-step}: & \hat{\boldsymbol{\theta}}^{(k)} = \arg\max_{\boldsymbol{\theta}}\left\{Q\left(\boldsymbol{\theta},\hat{\boldsymbol{\theta}}^{(k-1)}\right)\right\} \end{cases} \tag{31}$$

- GEM (Bayesian) algorithm:

$$\begin{cases} \text{E-step}: & Q\left(\boldsymbol{\theta},\hat{\boldsymbol{\theta}}^{(k)}\right) = \mathrm{E}_{p\left(\boldsymbol{f}|\boldsymbol{g},\hat{\boldsymbol{\theta}}^{(k)}\right)}\{\ln p(\boldsymbol{g},\boldsymbol{f}|\boldsymbol{\theta}) + \ln p(\boldsymbol{\theta})\} \\ \text{M-step}: & \hat{\boldsymbol{\theta}}^{(k)} = \arg\max_{\boldsymbol{\theta}}\left\{Q\left(\boldsymbol{\theta},\hat{\boldsymbol{\theta}}^{(k-1)}\right)\right\} \end{cases} \tag{32}$$

These methods can be summarized in the following scheme:

$$\boxed{p(\boldsymbol{f},\boldsymbol{\theta}|\boldsymbol{g})} \rightarrow \boxed{\text{EM, GEM}} \rightarrow \hat{\boldsymbol{\theta}} \rightarrow \boxed{p\left(\boldsymbol{f}|\hat{\boldsymbol{\theta}},\boldsymbol{g}\right)} \rightarrow \hat{\boldsymbol{f}}$$

5.4 Variational Bayesian approximation

VBA is a powerful approach to do approximate Bayesian computation. It starts by first obtaining the expression of the joint $p(f,\boldsymbol{\theta}|g)$ and then by approximating it with a simpler probability law $q(\boldsymbol{f},\boldsymbol{\theta}|g)$ which can be handled much easily for the computations. VBA can be summarized in the following steps:

- Approximate $p(\boldsymbol{f},\boldsymbol{\theta}|\boldsymbol{g})$ by $q(\boldsymbol{f},\boldsymbol{\theta}|\boldsymbol{g}) = q_1(\boldsymbol{f}|\boldsymbol{g})\, q_2(\boldsymbol{\theta}|\boldsymbol{g})$ and then continue computations.

- To do this approximation, we need a criterion to qualify the approximation. The standard criterion to measure the proximity of two probability laws p and q is the Kullback–Leibler (KL) criterion $\mathrm{KL}\,(q(\boldsymbol{f},\boldsymbol{\theta}|\boldsymbol{g}) : p(\boldsymbol{f},\boldsymbol{\theta}|\boldsymbol{g}))$.

- It is easy to show that:

$$\begin{aligned} \mathrm{KL}(q:p) &= \iint q \ln q/p = \iint q_1 q_2 \ln \frac{q_1 q_2}{p} \\ &= \int q_1 \ln q_1 + \int q_2 \ln q_2 - \iint q \ln p \\ &= -H(q_1) - H(q_2) - <\ln p>_q \end{aligned} \tag{33}$$

- Alternate optimization of KL($q_1 q_2$:p) with respect to q_1 and q_2 results to:

$$\begin{cases} q_1(\boldsymbol{f}) & \propto \exp\left[\langle \ln p(\boldsymbol{g},\boldsymbol{f},\boldsymbol{\theta};\mathcal{M})\rangle_{q_2(\boldsymbol{\theta})}\right] \\ q_2(\boldsymbol{\theta}) & \propto \exp\left[\langle \ln p(\boldsymbol{g},\boldsymbol{f},\boldsymbol{\theta};\mathcal{M})\rangle_{q_1(\boldsymbol{f})}\right] \end{cases} \tag{34}$$

As KL($q_1 q_2$:p) is convex as a function of q_1 and q_2, the algorithm converges (locally) to the optimum solution. At the end, we have the expressions of $q_1(\boldsymbol{f})$ and $q_2(\boldsymbol{\theta})$ which can, then, be used to infer on $\boldsymbol{f}$ and $\boldsymbol{\theta}$. VBA is summarized in the following scheme:

$$\boxed{p(\boldsymbol{f},\boldsymbol{\theta}|\boldsymbol{g})} \rightarrow \boxed{\begin{array}{c}\text{Variation}\\ \text{Bayesian}\\ \text{Approximation}\end{array}} \begin{array}{l} \rightarrow q_1(\boldsymbol{f}) \rightarrow \hat{\boldsymbol{f}} \\ \rightarrow q_2(\boldsymbol{\theta}) \rightarrow \hat{\boldsymbol{\theta}} \end{array}$$

In real applications, we choose parametric probability law for $q_1(\boldsymbol{f})$ $q_2(\boldsymbol{\theta})$, and so, the iterations will be done on the parameters. What is interesting is that, choosing appropriate parametric models for $q_1(\boldsymbol{f})$ and $q_2(\boldsymbol{\theta})$ we obtain either JMAP and GEM as special cases.

- Case 1: Deterministic or degenerate expressions $\rightarrow$ Joint MAP

$$\begin{cases} \hat{q}_1(\boldsymbol{f}|\tilde{\boldsymbol{f}}) & = \delta(\boldsymbol{f}-\tilde{\boldsymbol{f}}) \\ \hat{q}_2(\boldsymbol{\theta}|\tilde{\boldsymbol{\theta}}) & = \delta(\boldsymbol{\theta}-\tilde{\boldsymbol{\theta}}) \end{cases} \rightarrow \begin{cases} \tilde{\boldsymbol{f}} = \arg\max_{\boldsymbol{f}}\{p(\boldsymbol{f},\tilde{\boldsymbol{\theta}}|\boldsymbol{g};\mathcal{M})\} \\ \tilde{\boldsymbol{\theta}} = \arg\max_{\boldsymbol{\theta}}\{p(\tilde{\boldsymbol{f}},\boldsymbol{\theta}|\boldsymbol{g};\mathcal{M})\} \end{cases} \tag{35}$$

- Case 2: Degenerate expression for $\boldsymbol{\theta}$ and marginal expression for $\boldsymbol{f}$ $\rightarrow$ EM

$$\begin{cases} \hat{q}_1(\boldsymbol{f}) \propto p(\boldsymbol{f}|\boldsymbol{\theta},\boldsymbol{g}) \\ \hat{q}_2(\boldsymbol{\theta}|\tilde{\boldsymbol{\theta}}) = \delta(\boldsymbol{\theta}-\tilde{\boldsymbol{\theta}}) \end{cases} \rightarrow \begin{cases} Q(\boldsymbol{\theta},\tilde{\boldsymbol{\theta}}) & = \langle \ln p(\boldsymbol{f},\boldsymbol{\theta}|\boldsymbol{g};\mathcal{M})\rangle_{q_1(\boldsymbol{f}|\tilde{\boldsymbol{\theta}})} \\ \tilde{\boldsymbol{\theta}} & = \arg\max_{\boldsymbol{\theta}}\{Q(\boldsymbol{\theta}|\tilde{\boldsymbol{\theta}})\} \end{cases} \tag{36}$$

- Case 3: q_1 and q_2 are chosen proportional to the marginals $p(\boldsymbol{f}|\tilde{\boldsymbol{\theta}},\boldsymbol{g};\mathcal{M})$ and $p(\boldsymbol{\theta}|\tilde{\boldsymbol{f}},\boldsymbol{g};\mathcal{M})$. This is a very appropriate choice for inverse problems, in particular cases where we use the exponential families and conjugate priors.

$$\begin{cases} \hat{q}_1(\boldsymbol{f}) & \propto p(\boldsymbol{f}|\tilde{\boldsymbol{\theta}},\boldsymbol{g};\mathcal{M}) \\ \hat{q}_2(\boldsymbol{\theta}) & \propto p(\boldsymbol{\theta}|\tilde{\boldsymbol{f}},\boldsymbol{g};\mathcal{M}) \end{cases} \rightarrow \begin{cases} \text{Accounts for the uncertainties of} \\ \tilde{\boldsymbol{\theta}} \text{ for } \hat{\boldsymbol{f}} \text{ and vise versa.} \end{cases} \tag{37}$$

In the following schemes these three cases are illustrated for comparison.

- JMAP Alternate optimization Algorithm:

$$\begin{array}{ccccc} \boldsymbol{\theta}^{(0)} \rightarrow \tilde{\boldsymbol{\theta}} \rightarrow & \boxed{\tilde{\boldsymbol{f}} = \arg\max_{\boldsymbol{f}}\{p(\boldsymbol{f},\tilde{\boldsymbol{\theta}}|\boldsymbol{g})\}} & \rightarrow \tilde{\boldsymbol{f}} \rightarrow \hat{\boldsymbol{f}} \\ \uparrow & & \downarrow \\ \hat{\boldsymbol{\theta}} \leftarrow \tilde{\boldsymbol{\theta}} \leftarrow & \boxed{\tilde{\boldsymbol{\theta}} = \arg\max_{\boldsymbol{\theta}}\{p(\tilde{\boldsymbol{f}},\boldsymbol{\theta}|\boldsymbol{g})\}} & \leftarrow \tilde{\boldsymbol{f}} \end{array}$$

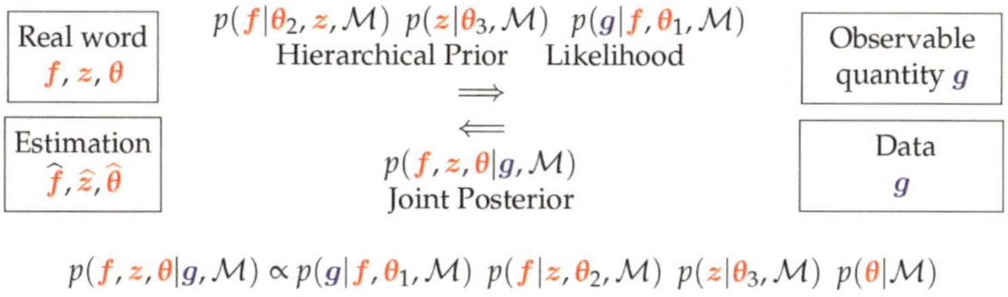

Figure 8.
Illustration of advanced Bayesian approach with hierarchical prior modeling with hidden variables.

- EM:

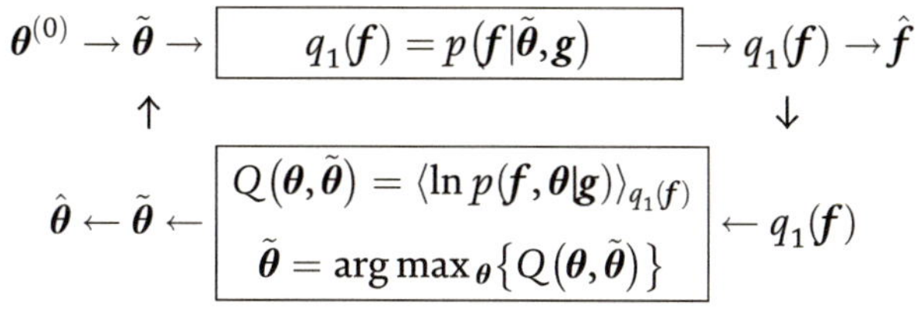

- VBA:

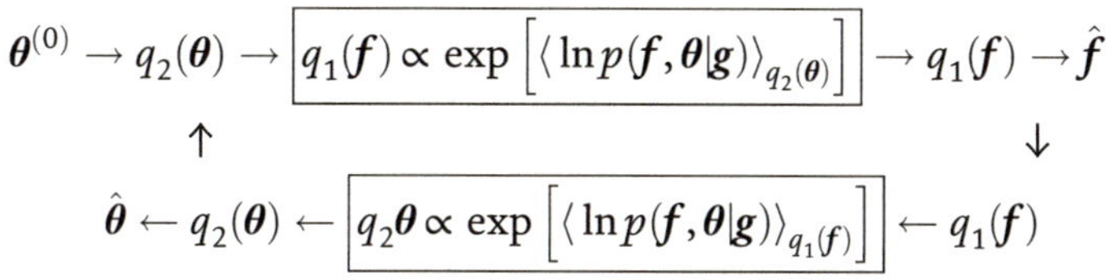

5.5 Hierarchical priors

One last extension is the case where $\boldsymbol{f}$, itself depends on another hidden variable $\boldsymbol{z}$. So that we have:

$$p(\boldsymbol{f},\boldsymbol{z},\boldsymbol{\theta}|\boldsymbol{g}) \propto p(\boldsymbol{g}|\boldsymbol{f},\boldsymbol{\theta}_1)p(\boldsymbol{f}|\boldsymbol{z},\boldsymbol{\theta}_2)p(\boldsymbol{z}|\boldsymbol{\theta}_3)p(\boldsymbol{\theta}), \tag{38}$$

where $\boldsymbol{\theta} = (\boldsymbol{\theta}_1, \boldsymbol{\theta}_2, \boldsymbol{\theta}_3)$. This situation is shown in **Figure 8**. Again, here, we may only be interested to $\boldsymbol{f}$ or $(\boldsymbol{f}, \boldsymbol{z})$ or to all the three variables $(\boldsymbol{f}, \boldsymbol{z}, \boldsymbol{\theta})$. Here too, we can either use methods of JMAP, marginalization or VBA to infer on these unknowns.

6. Linear forward models and Gaussian case

Linear models are of importance. Gaussian prior laws are the most common and the easiest ones to handle. Also, many non-linear problems can be approximated by equivalent linear ones. Linear models with Gaussian prior laws are the easiest and powerful tools for a great number of scientific problems. In this section, an overview and some main properties are given.

6.1 Simple supervised case

Let consider the linear forward model we considered in previous section

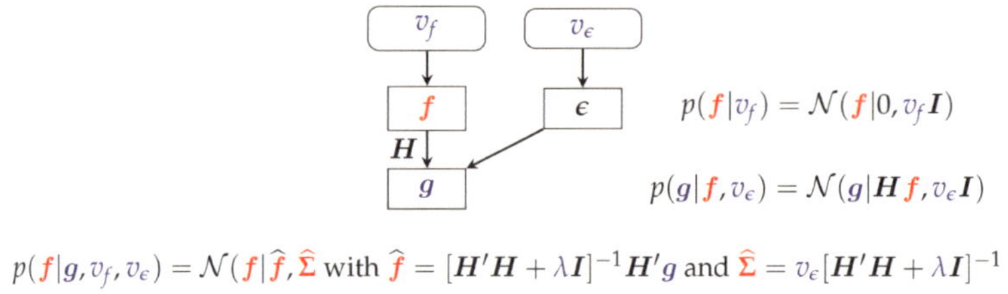

Figure 9.
Supervised linear Gaussian case.

$$\boldsymbol{g} = \boldsymbol{Hf} + \boldsymbol{\epsilon}, \tag{39}$$

and assign Gaussian laws to $\boldsymbol{\epsilon}$ and $\boldsymbol{f}$ which leads to:

$$\begin{cases} p(\boldsymbol{g}|\boldsymbol{f}) = \mathcal{N}(\boldsymbol{g}|\boldsymbol{Hf}, v_\epsilon \boldsymbol{I}) \propto \exp\left[-\frac{1}{2v_\epsilon}\|\boldsymbol{g} - \boldsymbol{Hf}\|^2\right] \\ p(\boldsymbol{f}) = \mathcal{N}(\boldsymbol{f}|0, v_f \boldsymbol{I}) \propto \exp\left[-\frac{1}{2v_f}\|\boldsymbol{f}\|^2\right] \end{cases} \tag{40}$$

Using these expressions, we get:

$$\begin{cases} p(\boldsymbol{f}|\boldsymbol{g}) \propto \exp\left[-\frac{1}{2v_\epsilon}\|\boldsymbol{g} - \boldsymbol{Hf}\|^2 - \frac{1}{2v_f}\|\boldsymbol{f}\|^2\right] \\ \qquad \propto \exp\left[-\frac{1}{2v_\epsilon}J(\boldsymbol{f})\right] \text{with } J(\boldsymbol{f}) = \|\boldsymbol{g} - \boldsymbol{Hf}\|^2 + \lambda\|\boldsymbol{f}\|^2, \quad \lambda = \frac{v_\epsilon}{v_f} \end{cases} \tag{41}$$

which can be summarized as:

$$p(\boldsymbol{f}|\boldsymbol{g}) = \mathcal{N}\left(\boldsymbol{f}|\hat{\boldsymbol{f}}, \hat{\Sigma}\right) \text{ with } \hat{\boldsymbol{f}} = [\boldsymbol{H}'\boldsymbol{H} + \lambda I]^{-1}\boldsymbol{H}'\boldsymbol{g} \text{ and } \hat{\Sigma} = v_\epsilon[\boldsymbol{H}'\boldsymbol{H} + \lambda I]^{-1}, \tag{42}$$

where $\lambda = \frac{v_\epsilon}{v_f}$. This case is summarized in **Figure 9**.

This is the simplest case where we know exactly the expression of the posterior law and all the computations can be done explicitly. However, for great dimensional problems, where the vectors $\boldsymbol{f}$ and $\boldsymbol{g}$ are very great dimensional, we may even not be able to keep in memory the matrix $\boldsymbol{H}$ and surely not be able to compute the inverse of the matrix $[\boldsymbol{H}'\boldsymbol{H} + \lambda\boldsymbol{I}]$. In Section 9 on Bayesian computation, We will see how to do these computations.

6.2 Unsupervised case or hyperparameter estimation

In the previous section, we considered the linear models with Gaussian priors with known parameters v_ϵ and v_f. In many practical situations these parameters are not known, and we want to estimate them too. For this, we can assign them too prior laws. As the variances are positive quantities and using the concept of conjugate priors, we can assign then Inverse Gamma priors:

$$\begin{cases} p(v_\epsilon) = \mathcal{IG}(v_f|\alpha_{\epsilon_0}, \beta_{\epsilon_0}) \\ p(v_f) = \mathcal{IG}\left(v_f|\alpha_{f_0}, \beta_{f_0}\right) \end{cases} \tag{43}$$

and using the likelihood $p(\boldsymbol{g}|\boldsymbol{f}, v_\epsilon) = \mathcal{N}(\boldsymbol{g}|\boldsymbol{Hf}, v_\epsilon \boldsymbol{I})$ and the prior $p(\boldsymbol{f}|v_f) = \mathcal{N}(\boldsymbol{f}|0, v_f\boldsymbol{I})$, we can easily obtain the expressions of the following conditional posterior laws:

$$\begin{cases} p(\boldsymbol{f}|\boldsymbol{g}, \hat{v}_\epsilon, \hat{v}_f) = \mathcal{N}\left(\boldsymbol{f}|\hat{\boldsymbol{f}}, \hat{\Sigma}\right) \quad \text{with :} \\ \hat{\boldsymbol{f}} = \left[\boldsymbol{H}^t\boldsymbol{H} + \hat{\lambda}\boldsymbol{I}\right]^{-1}\boldsymbol{H}^t\boldsymbol{g} \\ \hat{\Sigma} = \hat{v}_\epsilon\left[\boldsymbol{H}^t\boldsymbol{H} + \hat{\lambda}\boldsymbol{I}\right]^{-1}, \quad \hat{\lambda} = \hat{v}_\epsilon/\hat{v}_f \end{cases} \tag{44}$$

and

$$\begin{cases} p(v_\epsilon|\boldsymbol{g},\boldsymbol{f}) = \mathcal{IG}(v_\epsilon|\tilde{\alpha}_\epsilon, \tilde{\beta}\epsilon) \\ p(v_f|\boldsymbol{g},\boldsymbol{f}) = \mathcal{IG}\left(v_f|\tilde{\alpha}_f, \tilde{\beta}_f\right) \end{cases} \tag{45}$$

where all the details and in particular the expressions for $\tilde{\alpha}_\epsilon, \tilde{\beta}_\epsilon, \tilde{\alpha}_f, \tilde{\beta}_f$ can be found in [10].

As we can see, the expressions of $\hat{\boldsymbol{f}}$ and $\hat{\boldsymbol{\Sigma}}$ are the same as in the previous case, except that the values of $\hat{v}_\epsilon, \hat{v}_f$ and $\hat{\lambda}$ have to be updated. They are obtained from the conditionals $p(v_\epsilon|\boldsymbol{g},\boldsymbol{f})$ and $p(v_f|\boldsymbol{g},\boldsymbol{f})$ which depend on $\boldsymbol{f}$. This shows that we can propose an iterative algorithm in two steps: Determine the expression of $p(\boldsymbol{f}|\boldsymbol{g}, \hat{v}_\epsilon, \hat{v}_f)$ and using the values of in the previous iteration, we can propose an estimate for $\boldsymbol{f}$, and then, using $p(v_\epsilon|\boldsymbol{g},\boldsymbol{f})$ and $p(v_f|\boldsymbol{g},\boldsymbol{f})$, we can give estimates for $\hat{v}_\epsilon$ and $\hat{v}_f$ which can again be used in the first step. It is interesting to know that all the three approaches of JMAP, GEM and VBA for this cas follow exactly this same iterative algorithm. The only differences will be in the update values of $\tilde{\alpha}_\epsilon, \tilde{\beta}_\epsilon, \tilde{\alpha}_f, \tilde{\beta}_f$ and the choice of the estimators (MAP or PM) of $\hat{v}_\epsilon$ and $\hat{v}_f$.

This case is also summarized in **Figure 10**.

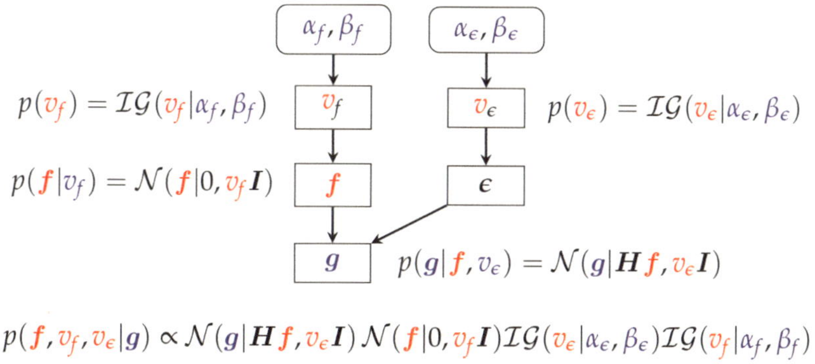

Figure 10.
Bayesian inference scheme in linear systems and Gaussian priors. The posterior is also Gaussian, and all the computations can be done analytically.

7. Non-Gaussian priors

Very often, assuming that the noise is Gaussian is valid in many applications, but a Gaussian prior may not be adequate. Thus, the case of Non-Gaussian priors is of great importance. A very well known example is the case of Generalized Gaussian:

$$p(\boldsymbol{f}) \propto \exp\left[-\gamma \sum_j |\boldsymbol{f}_j|^{\beta}\right]. \tag{46}$$

The case of $\beta = 2$ is the Gaussian case, $\beta > 2$ gives the *Super-Gaussian* and $\beta < 1$ is called *Sub-Gaussian*. Its particular case $\beta = 1$ results to what is called *Double Exponential (DE)* prior law:

$$p(\boldsymbol{f}) \propto \exp\left[-\gamma \sum_j |f_j|\right] \propto \exp\left[-\gamma \|\boldsymbol{f}\|_1\right] \tag{47}$$

which, when using with a Gaussian likelihood, results to:

$$\begin{cases} p(\boldsymbol{f}|\boldsymbol{g}) & \propto \exp\left[-\frac{1}{2v_\epsilon} J(\boldsymbol{f})\right] \quad \text{with} \\ J(\boldsymbol{f}) & = \frac{1}{2}\|\boldsymbol{g} - \boldsymbol{H} \quad \boldsymbol{f}\|^2 + \lambda \|\boldsymbol{f}\|_1, \quad \lambda = \gamma v_\epsilon. \end{cases} \tag{48}$$

From this, we can see that the computation of MAP solution needs an appropriate optimization algorithm and the computations of the Posterior Mean (PM) or Posterior Covariance (PCov) or any other expectations become more difficult. However, as we will see later, VBA can be used to do approximate computations.

Another example is the Total Variation (TV) regularization method [11–14] which can be interpreted as choosing the prior

$$p(\boldsymbol{f}) \propto \exp\left[-\gamma \sum_j |f_j - f_{j-1}|\right] \propto \exp\left[-\gamma \|\boldsymbol{D}\boldsymbol{f}\|_1\right] \tag{49}$$

where $\boldsymbol{D}$ is the first order difference matrix.

This prior with a Gaussian model for noise results to:

$$\begin{cases} p(\boldsymbol{f}|\boldsymbol{g}) & \propto \exp\left[-\frac{1}{2v\epsilon} J(\boldsymbol{f})\right] \quad \text{with} \\ J(\boldsymbol{f}) & = \frac{1}{2}\|\boldsymbol{g} - \boldsymbol{H}\boldsymbol{f}\|^2 + \lambda \|\boldsymbol{D}\boldsymbol{f}\|_1, \quad \lambda = \gamma v_\epsilon. \end{cases} \tag{50}$$

One last example is using the Cauchy or more generally the Student-t distribution as the prior:

$$p(\boldsymbol{f}) \propto \exp\left[-\gamma \sum_j \ln\left(1 + f_j^2\right)^{\nu/2}\right] \tag{51}$$

which results to:

$$\begin{cases} p(\boldsymbol{f}|\boldsymbol{g}) & \propto \exp\left[-\frac{1}{2v_\epsilon} J(\boldsymbol{f})\right] \quad \text{with} \\ J(\boldsymbol{f}) & = \frac{1}{2}\|\boldsymbol{g} - \boldsymbol{H}\boldsymbol{f}\|^2 + \lambda \sum_j \ln\left(1 + f_j^2\right)^{\nu/2}, \quad \lambda = \gamma v_\epsilon. \end{cases} \tag{52}$$

These three examples are of great importance. They have been used in the framework of MAP estimation and thus the optimization of the criteria $J(f)$ for many linear inverse problems. However, the computation of other Bayesian estimators and uncertainty quantification (UQ) need again specific approximate solutions.

8. Hierarchical prior models

Even if simple Gaussian and non-Gaussian priors used in previous sections are of great importance and use in many applications, still they have, in many cases, limitations. For example, when we know that the signals have impulsive shapes or discontinuous or are piecewise continuous. The same limitations when we know, for example, that the images are composed of homogeneous regions with specified contours, or even, that the object under the test is composed of a limited number of homogeneous materials. Hierarchical models push farther these limitations of simple prior models. In the following, we consider three families of such hierarchical models: Sparsity aware models, Scaled Mixture models and Gauss-Markov-Potts models [10, 15–17].

8.1 Sparsity awarded hierarchical models

An easy way to consider the hierarchical sparsity awarded priors is to introduce a hidden variable, $\boldsymbol{z}$ and so consider the following Forward and prior models:

$$\begin{cases} \boldsymbol{g} = \boldsymbol{H}\boldsymbol{f} + \boldsymbol{\epsilon}, \\ \boldsymbol{f} = \boldsymbol{D}\boldsymbol{z} + \boldsymbol{\zeta}, \quad \boldsymbol{z} \text{ sparse modeled by Double Exp (DE)} \end{cases} \tag{53}$$

with

$$\begin{cases} p(\boldsymbol{g}|\boldsymbol{f}) = \mathcal{N}(\boldsymbol{g}|\boldsymbol{H}\boldsymbol{f}, v_\epsilon \boldsymbol{I}) \\ p(\boldsymbol{f}|\boldsymbol{z}) = \mathcal{N}(\boldsymbol{f}|\boldsymbol{D}\boldsymbol{z}, v_\xi \boldsymbol{I}) \rightarrow \\ p(\boldsymbol{z}) = \mathcal{DE}(\boldsymbol{f}|\gamma) \propto \exp\left[-\gamma\|\boldsymbol{z}\|_1\right] \end{cases} \tag{54}$$

Then, we have to find the expression of the joint posterior law $p(\boldsymbol{f}, \boldsymbol{z}|\boldsymbol{g})$:

$$\begin{cases} p(\boldsymbol{f}, \boldsymbol{z}|\boldsymbol{g}) \propto \exp\left[-J(\boldsymbol{f}, \boldsymbol{z})\right] \quad \text{with} \\ J(\boldsymbol{f}, \boldsymbol{z}) = \dfrac{1}{2v_\epsilon}\|\boldsymbol{g} - \boldsymbol{H}\boldsymbol{f}\|_2^2 + \dfrac{1}{2v_\xi}\|\boldsymbol{f} - \boldsymbol{D}\boldsymbol{z}\|_2^2 + \gamma\|\boldsymbol{z}\|_1 \end{cases} \tag{55}$$

from which we can infer on $\boldsymbol{f}$ and $\boldsymbol{z}$ [10, 16, 18–22].
For the unsupervised case, we can add the appropriate priors:

$$\begin{cases} p(\gamma) = \mathcal{IG}\left(\gamma|\alpha_{\gamma_z}, \beta_{\gamma_z}\right) \\ p(v_\epsilon) = \mathcal{IG}\left(v_\epsilon|\alpha_{\epsilon_0}, \beta_{\epsilon_0}\right) \\ p(v_\xi) = \mathcal{IG}\left(v_\xi|\alpha_{\xi_z}, \beta_{\xi_z}\right) \end{cases} \tag{56}$$

and thus obtain:

$$\begin{cases} p(\boldsymbol{f}, \boldsymbol{z}, \gamma, v_\epsilon, v_\xi | \boldsymbol{g}) \propto \exp\left[-J(\boldsymbol{f}, \boldsymbol{z}, \gamma, v_\epsilon, v_\xi)\right] \quad \text{with} \\ J(\boldsymbol{f}, \boldsymbol{z}, v_\epsilon, v_\xi, \gamma) = \frac{1}{2v_\epsilon}\|\boldsymbol{g} - \boldsymbol{Hf}\|_2^2 + \frac{1}{2v_\xi}\|\boldsymbol{f} - \boldsymbol{Dz}\|_2^2 + \gamma\|\boldsymbol{z}\|_1 + \\ \qquad (\alpha_{\gamma_z} + n/2)\ln\gamma + \beta_{\gamma_z}/\gamma + \\ \qquad (\alpha_{\epsilon 0} + m/2)\ln v_\epsilon + \beta_{\epsilon 0}/v_\epsilon + \\ \qquad (\alpha_{\xi_z} + n/2)\ln v_\xi + \beta_{\xi_z}/v_\xi \end{cases} \tag{57}$$

It is interesting to note that the alternate optimization of this criterion gives the ADMM like algorithms [23–25] with the main advantage that here we have direct updates of the hyperparameters.

8.2 Scaled mixture models

Scaled Gaussian Mixture (SGM) models have been used in many applications to model rare events by their heavier tails with respect to Gaussian. They are also used in sparse signals modeling. A general SGM is defined as follows:

$$\mathcal{S}(f) = \int \mathcal{N}(f|0, v)\, p_m(v|\boldsymbol{\theta})\, \mathrm{d}v \tag{58}$$

where the variance of the Gaussian model $\mathcal{N}(f|0, v)$ is assumed to follow the mixing probability law $p_m(v|\boldsymbol{\theta})$. Between many possibilities for this mixing pdf is Inverse-Gamma which results to Student-t:

$$\mathcal{S}(f|\nu) = \int \mathcal{N}(f|0, v)\mathcal{IG}(v|\nu, \nu)\, \mathrm{d}v \tag{59}$$

which have been extended to more general case:

$$\mathcal{S}(f|\alpha, \beta) = \int \mathcal{N}(f|0, v)\mathcal{IG}(v|\alpha, \beta)\, \mathrm{d}v \tag{60}$$

This pdf models have been used with success in many developments in Bayesian approach for inverse problems by:

$$p(\boldsymbol{f}|\alpha, \beta) = \prod_j \int \mathcal{N}\left(f_j|0, v_j\right)\mathcal{IG}\left(v_j|\alpha, \beta\right)\, \mathrm{d}v_j \tag{61}$$

or

$$p(\boldsymbol{f}|\alpha, \beta) = \int \mathcal{N}(\boldsymbol{f}|0, v\Sigma)\mathcal{IG}(v|\alpha, \beta)\, \mathrm{d}v \tag{62}$$

Scaled Gaussian mixture models have been used extensively for modeling sparse signals. However, it happens very often that the signals or images are not sparse directly, but their gradients are, or more generally in a transformed domain such as Fourier or Wavelet domains. We have used these models extensively in hierarchical way:

$$\begin{cases} \boldsymbol{g} = \boldsymbol{Hf} + \boldsymbol{\epsilon}, \\ \boldsymbol{f} = \boldsymbol{Dz} + \zeta, \quad \boldsymbol{z} \text{ sparse Student} \rightarrow \begin{cases} p\left(z_j|v_{z_j}\right) = \mathcal{N}\left(z_j|0, v_{z_j}\right), \\ p\left(v_{z_j}\right) = \mathcal{IG}\left(v_{z_j}|\alpha_{z0}, \beta_{z0}\right) \end{cases} \end{cases} \tag{63}$$

where $\boldsymbol{D}$ represents any linear transformations and $\boldsymbol{D}^{-1}$ applied of $\boldsymbol{f}$ transforms it to a sparse vector $\boldsymbol{z}$.

The whole relations of the likelihood and priors are summarized in below:

$$\begin{cases} p(\boldsymbol{g}|\boldsymbol{f}) = \mathcal{N}(\boldsymbol{g}|\boldsymbol{H}\boldsymbol{f}, v_\epsilon \boldsymbol{I}) \\ p(\boldsymbol{f}|\boldsymbol{z}) = \mathcal{N}(\boldsymbol{f}|\boldsymbol{D}\boldsymbol{z}, v_\xi \boldsymbol{I}) \\ p(\boldsymbol{z}|v_{\boldsymbol{z}}) = \mathcal{N}(\boldsymbol{z}|0, V_{\boldsymbol{z}}) \\ p(v_{\boldsymbol{z}}) = \prod_j \mathcal{IG}\left(v_{\boldsymbol{z}_j}|\alpha_{\boldsymbol{z}0}, \beta_{\boldsymbol{z}0}\right) \\ p(v_\epsilon) = \mathcal{IG}(v_\epsilon|\alpha_{\epsilon 0}, \beta_{\epsilon 0}) \\ p(v_\xi) = \mathcal{IG}\left(v_\xi|\alpha_{\xi_{\boldsymbol{z}}}, \beta_{\xi_{\boldsymbol{z}}}\right) \end{cases} \tag{64}$$

and the corresponding joint posterior of all the unknowns writes:

$$\begin{cases} p(\boldsymbol{f}, \boldsymbol{z}, \boldsymbol{v}_{\boldsymbol{z}}, v_\epsilon, v_\xi|\boldsymbol{g}) \propto \exp\left[-J(\boldsymbol{f}, \boldsymbol{z}, \boldsymbol{v}_{\boldsymbol{z}}, v_\epsilon, v_\xi)\right] \\ J(\boldsymbol{f}, \boldsymbol{z}, \boldsymbol{v}_{\boldsymbol{z}}, v_\epsilon, v_\xi) = \dfrac{1}{2v_\epsilon}\|\boldsymbol{g} - \boldsymbol{H}\boldsymbol{f}\|_2^2 + \dfrac{1}{2v_\xi}\|\boldsymbol{f} - \boldsymbol{D}\boldsymbol{z}\|_2^2 + \left\|\boldsymbol{V}_{\boldsymbol{z}}^{\frac{-1}{2}}\boldsymbol{z}\right\|_2^2 + \\ \qquad \sum_j (\alpha_{\boldsymbol{z}_0} + 1)\ln v_{\boldsymbol{z}_j} + \beta_{\boldsymbol{z}_0}/v_{\boldsymbol{z}_j} + \\ \qquad (\alpha_{\epsilon_0} + m/2)\ln v_\epsilon + \beta_{\epsilon_0}/v_\epsilon + (\alpha_{\xi_{\boldsymbol{z}}} + n/2)\ln v_\xi + \beta_{\xi_{\boldsymbol{z}}}/v_\xi \end{cases} \tag{65}$$

Looking at this expression, we see that we have:

- Quadratic optimization with respect to $\boldsymbol{f}$ and $\boldsymbol{z}$;
- Direct analytical expressions for the updates of the hyperparameters v_ϵ and v_ξ;
- Possibility to compute posterior mean and quantify uncertainties analytically via VBA.

A final case we consider is the case of Non-stationary noise and sparsity enforcing prior in the same framework.

$$\begin{cases} \boldsymbol{g} = \boldsymbol{H}\boldsymbol{f} + \boldsymbol{\epsilon}, \quad \boldsymbol{\epsilon} \text{ non stationary} \rightarrow \begin{cases} p(\epsilon_i|v_{\epsilon_i}) = \mathcal{N}(\epsilon_i|0, v_{\epsilon_i}), \\ p(v_{\epsilon_i}) = \mathcal{IG}(v_{\epsilon_i}|\alpha_{\epsilon 0}, \beta_{\epsilon 0}) \end{cases} \\ \boldsymbol{f} = \boldsymbol{D}\boldsymbol{z} + \boldsymbol{\zeta}, \quad \boldsymbol{z} \text{ sparse Student} \rightarrow \begin{cases} p\left(z_j|v_{z_j}\right) = \mathcal{N}\left(z_j|0, v_{z_j}\right), \\ p\left(v_{z_j}\right) = \mathcal{IG}\left(v_{z_j}|\alpha_{z0}, \beta_{z0}\right) \end{cases} \end{cases} \tag{66}$$

Again here, all the expressions of likelihood and priors can be summarized as follows:

$$\begin{cases} p(\boldsymbol{g}|\boldsymbol{f}) = \mathcal{N}(\boldsymbol{g}|\boldsymbol{H}\boldsymbol{f}, V_\epsilon) \\ p(\boldsymbol{f}|\boldsymbol{z}) = \mathcal{N}(\boldsymbol{f}|D\boldsymbol{z}, v_\xi I) \\ p(\boldsymbol{z}|\boldsymbol{v}_{\boldsymbol{z}}) = \mathcal{N}(\boldsymbol{z}|0, V_{\boldsymbol{z}}) \\ p(\boldsymbol{v}_{\boldsymbol{z}}) = \prod_j \mathcal{IG}\left(v_{\boldsymbol{z}_j}|\alpha_{\boldsymbol{z}_0}, \beta_{\boldsymbol{z}0}\right) \\ p(\boldsymbol{v}_\epsilon) = \prod_i \mathcal{IG}(v_{\epsilon_i}|\alpha_{\epsilon 0}, \beta_{\epsilon 0}) \\ p(v_\xi) = \mathcal{IG}\left(v_\xi|\alpha_{\xi_{\boldsymbol{z}}}, \beta_{\xi_{\boldsymbol{z}}}\right) \end{cases} \tag{67}$$

and the joint posterior of all the unknowns become:

$$\begin{cases} p(\boldsymbol{f},\boldsymbol{z},\boldsymbol{v}_{\boldsymbol{z}},\boldsymbol{v}_\epsilon,v_\xi|\boldsymbol{g}) \propto \exp\left[-J(\boldsymbol{f},\boldsymbol{z},\boldsymbol{v}_{\boldsymbol{z}},\boldsymbol{v}_\epsilon,v_\xi)\right] \\ J(\boldsymbol{f},\boldsymbol{z},\boldsymbol{v}_{\boldsymbol{z}},v_\epsilon,v_\xi) = \left\|\boldsymbol{V}_\epsilon^{\frac{-1}{2}}(\boldsymbol{g}-\boldsymbol{H}\boldsymbol{f})\right\|_2^2 + \frac{1}{2v_\xi}\|\boldsymbol{f}-\boldsymbol{D}\boldsymbol{z}\|_2^2 + \left\|\boldsymbol{V}_{\boldsymbol{z}}^{\frac{-1}{2}}\boldsymbol{z}\right\|_2^2 + \\ \qquad \sum_j(\alpha_{\boldsymbol{z}_0}+1)\ln v_{\boldsymbol{z}_j} + \beta_{\boldsymbol{z}_0}/v_{\boldsymbol{z}_j} + \\ \qquad \sum_j(\alpha_{\epsilon_0}+1)\ln v_{\epsilon_i} + \beta_{\epsilon_0}/v_{\epsilon_i} + \\ \qquad (\alpha_{\xi_{\boldsymbol{z}}}+n/2)\ln v_\xi + \beta_{\xi_{\boldsymbol{z}}}/v_\xi \end{cases} \tag{68}$$

The following scheme shows graphically this case.

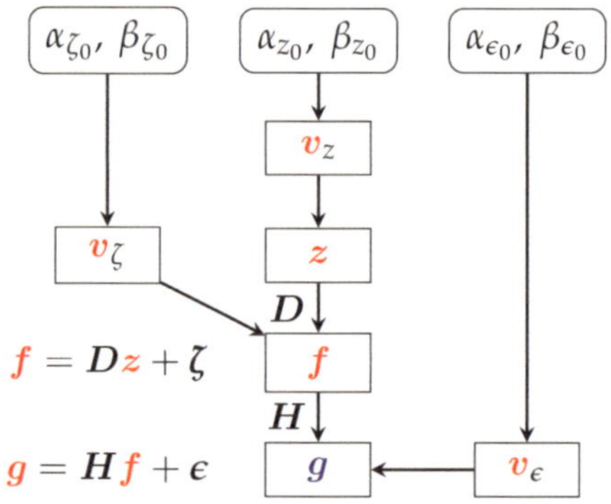

$$p(\boldsymbol{f},\boldsymbol{z},v_z,v_\epsilon,v_\xi|\boldsymbol{g}) \propto \mathcal{N}(\boldsymbol{g}|\boldsymbol{H}\boldsymbol{f},v_\epsilon\boldsymbol{I})\mathcal{N}(\boldsymbol{f}|\boldsymbol{D}\boldsymbol{z},v_\xi\boldsymbol{I})\mathcal{N}(\boldsymbol{z}|0,\boldsymbol{V}_z)$$
$$\mathcal{IG}(v_z|\alpha_{z_0},\beta_{z_0})\mathcal{IG}(v_\epsilon|\alpha_{\epsilon_0},\beta_{\epsilon_0})\mathcal{IG}(v_\xi|\alpha_{\xi_z},\beta_{\xi_z})$$

8.3 A four level hierarchical model

To account separately for the measurement and forward modeling error, a more detailed and four level hierarchical model has been proposed:

$$\begin{cases} \boldsymbol{g} = \boldsymbol{g}_0 + \boldsymbol{\epsilon}, & \text{measurement error} \\ \boldsymbol{g}_0 = \boldsymbol{H}\boldsymbol{f} + \xi, & \text{modeling error} \\ \boldsymbol{f} = \boldsymbol{D}\boldsymbol{z} + \boldsymbol{\zeta}, & \text{Prior model} \end{cases} \tag{69}$$

which accounts for two terms of errors (variable splitting) and Sparsity enforcing in Transformed domain prior: $\boldsymbol{f} = \boldsymbol{D}\boldsymbol{z} + \boldsymbol{\zeta}$ with $\boldsymbol{z}$ sparse, modeled itself by Normal-IG.

This model is presented graphically here.

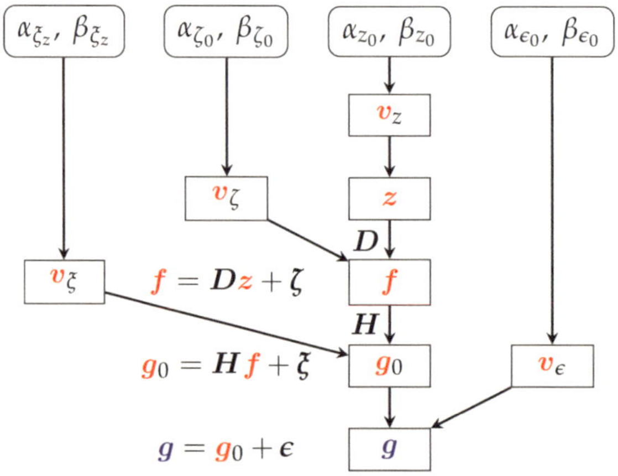

In this model, there are three error terms: $\boldsymbol{\epsilon}$ the observation error, $\boldsymbol{\zeta}$ the forward modeling error and $\boldsymbol{\zeta}$ the transform domain modeling error. These are detailed in the following:

- $\boldsymbol{g} = \boldsymbol{g}_0 + \boldsymbol{\epsilon},: \boldsymbol{\epsilon}$ is assumed to be Gaussian:

$$p(\boldsymbol{g}|\boldsymbol{g}_0, v_\epsilon) = \mathcal{N}(\boldsymbol{g}|\boldsymbol{g}_0, v_\epsilon I), p(v_\epsilon) = \mathcal{IG}(v_\epsilon|\alpha_{\epsilon_0}, \beta_{\epsilon_0}),$$

- $\boldsymbol{g}_0 = \boldsymbol{H}\boldsymbol{f} + \xi, \xi$ is assumed to be Student-t:

$$\begin{cases} p(\boldsymbol{g}_0|\boldsymbol{f}, v_\xi) = \mathcal{N}(\boldsymbol{g}_0|\boldsymbol{H}\boldsymbol{f}, \boldsymbol{V}_\xi), \boldsymbol{V}_\xi = \text{diag}[v_\xi], \\ p(v_\xi) = \Pi_{i=1}^{M} p\left(v_{\xi_i}\right) = \Pi_{i=1}^{M} \mathcal{IG}\left(v_{\xi_i}|\alpha_{\xi_{\boldsymbol{z}}}, \beta_{\xi_{\boldsymbol{z}}}\right), \end{cases}$$

- $\boldsymbol{f} = \boldsymbol{D}\boldsymbol{z} + \boldsymbol{\zeta}, \boldsymbol{\zeta}$ is assumed to be Gaussian:

$$p(\boldsymbol{f}|\boldsymbol{z}, v_\zeta) = \mathcal{N}(\boldsymbol{f}|\boldsymbol{D}\boldsymbol{z}, v_\zeta I), p(v_\zeta) = \mathcal{IG}\left(v_\zeta|\alpha_{\zeta_0}, \beta_{\zeta_0}\right),$$

- $\boldsymbol{z}$ is assumed to be sparse and thus modeled via Normal-IG:

$$\begin{cases} p(\boldsymbol{z}|\boldsymbol{v}_{\boldsymbol{z}}) = \mathcal{N}(\boldsymbol{z}|0, \boldsymbol{V}_{\boldsymbol{z}}), \boldsymbol{V}_{\boldsymbol{z}} = \text{diag}[\boldsymbol{v}_{\boldsymbol{z}}] \\ p(\boldsymbol{v}_{\boldsymbol{z}}) = \Pi_{j=1}^{N} p\left(\boldsymbol{v}_{\boldsymbol{z}_j}\right) = \Pi_{j=1}^{N} \mathcal{IG}\left(\boldsymbol{v}_{\boldsymbol{z}_j}|\alpha_{\boldsymbol{z}_0}, \beta_{\boldsymbol{z}_0}\right) \end{cases}$$

which results in:

$$p(\boldsymbol{f}, \boldsymbol{g}_0, \boldsymbol{z}, \boldsymbol{v}_\epsilon, \boldsymbol{v}_\xi, \boldsymbol{v}_{\boldsymbol{z}}|\boldsymbol{g}) \propto \exp\left[-J(\boldsymbol{f}, \boldsymbol{g}_0, \boldsymbol{z}, \boldsymbol{v}_\epsilon, \boldsymbol{v}_\xi, \boldsymbol{v}_{\boldsymbol{z}})\right] \tag{70}$$

with

$$\begin{aligned} J(\boldsymbol{f}, \boldsymbol{g}_0, \boldsymbol{z}, \boldsymbol{v}_\epsilon, \boldsymbol{v}_\xi, \boldsymbol{v}_{\boldsymbol{z}}) = & \frac{1}{2\boldsymbol{v}_\epsilon}\|\boldsymbol{g} - \boldsymbol{g}_0\|_2^2 + (\alpha_{\epsilon_0} + 1)\ln \boldsymbol{v}_\epsilon + \frac{\beta_{\epsilon_0}}{\boldsymbol{v}_\epsilon} \\ & + \frac{1}{2}\left\|\boldsymbol{V}_\xi^{-1/2}(\boldsymbol{g}_0 - \boldsymbol{H}\boldsymbol{f})\right\|_2^2 + \sum_{i=1}^{M}\left[(\alpha_{\xi_{\boldsymbol{z}}} + 1)\ln \boldsymbol{v}_{\xi_i} + \frac{\beta_{\xi_{\boldsymbol{z}}}}{\boldsymbol{v}_{\xi_i}}\right] \\ & + \frac{1}{2\boldsymbol{v}_\zeta}\|\boldsymbol{f} - \boldsymbol{D}\boldsymbol{z}\|_2^2 + (\alpha_{\zeta_0} + 1)\ln \boldsymbol{v}_\xi + \frac{\beta_{\zeta_0}}{\boldsymbol{v}_\xi} \\ & + \frac{1}{2}\|\boldsymbol{V}_{\boldsymbol{z}}^{-1/2}\boldsymbol{z}\|_2^2 + \sum_{j=1}^{N}\left[(\alpha_{\boldsymbol{z}_0} + 1)\ln \boldsymbol{v}_{\boldsymbol{z}_j} + \frac{\beta_{\boldsymbol{z}_0}}{\boldsymbol{v}_{\boldsymbol{z}_j}}\right] \end{aligned} \tag{71}$$

Using then the JMAP approach with an alternate optimization strategy needs the following optimization steps:

- with respect to $\boldsymbol{f}$: $J(\boldsymbol{f}) = \frac{1}{2}\left\|\boldsymbol{V}_\xi^{-1/2}(\boldsymbol{g}_0 - \boldsymbol{H}\boldsymbol{f})\right\|_2^2 + \frac{1}{2\boldsymbol{v}_\zeta}\|\boldsymbol{f} - \boldsymbol{D}\boldsymbol{z}\|_2^2$

- with respect to $\boldsymbol{g}_0$: $J(\boldsymbol{g}_0) = \frac{1}{2\boldsymbol{v}_\epsilon}\|\boldsymbol{g} - \boldsymbol{g}_0\|_2^2 + \frac{1}{2}\left\|\boldsymbol{V}_\xi^{-1/2}(\boldsymbol{g}_0 - \boldsymbol{H}\boldsymbol{f})\right\|_2^2$

- with respect to $\boldsymbol{z}$: $J(\boldsymbol{z}) = \frac{1}{2\boldsymbol{v}_\zeta}\|\boldsymbol{f} - \boldsymbol{D}\boldsymbol{z}\|_2^2 + \frac{1}{2}\|\boldsymbol{V}_{\boldsymbol{z}}^{-1/2}\boldsymbol{z}\|_2^2$

- with respect to $\boldsymbol{v}_\epsilon : J(\boldsymbol{v}_\epsilon) = \frac{1}{2\boldsymbol{v}_\epsilon} \|\boldsymbol{g} - \boldsymbol{g}_0\|_2^2 + (\alpha_{\epsilon_0} + 1) \ln \boldsymbol{v}_\epsilon + \frac{\beta_{\epsilon_0}}{\boldsymbol{v}_\epsilon}$

- with respect to $\boldsymbol{v}_{\xi_i} : J(\boldsymbol{v}_{\xi_i}) = \frac{1}{2} \left\| \boldsymbol{V}_\xi^{-1/2} (\boldsymbol{g}_0 - \boldsymbol{H}\boldsymbol{f}) \right\|_2^2 + \sum_{i=1}^M \left[(\alpha_{\xi_{\boldsymbol{z}}} + 1) \ln \boldsymbol{v}_{\xi_i} + \frac{\beta_{\xi_{\boldsymbol{z}}}}{\boldsymbol{v}_{\xi_i}} \right]$

- with respect to $v_\zeta : J(v_\zeta) = \frac{1}{2v_\zeta} \|\boldsymbol{f} - \boldsymbol{D}\boldsymbol{z}\|_2^2 + (\alpha_{\zeta_0} + 1) \ln v_\xi + \frac{\beta_{\zeta_0}}{v_\xi}$

- to $v_{\boldsymbol{z}_j} : J\left(v_{\boldsymbol{z}_j}\right) = \frac{1}{2} \|\boldsymbol{V}_{\boldsymbol{z}}^{-1/2} \boldsymbol{z}\|_2^2 + \sum_{j=1}^N \left[(\alpha_{\boldsymbol{z}_0} + 1) \ln v_{\boldsymbol{z}_j} + \frac{\beta_{\boldsymbol{z}_0}}{v_{\boldsymbol{z}_j}} \right]$

This approach has the following main advantages and limitations.
Advantages:

- All the optimization are either quadratic or explicit
- Quadratic optimizations can be done efficiently
- For great dimensional problems, the needed operators $\boldsymbol{H}$, $\boldsymbol{H}'$, $\boldsymbol{D}$ and $\boldsymbol{D}'$ can be implemented on GPU
- For Computed Tomography, efficient GPU implementation of these operators have been done in our group for 2D and 3D CT.

Limitations:

- Huge amount of memory is needed for $\boldsymbol{f}$, $\boldsymbol{g}_0$, v_ξ and $\boldsymbol{v}_z$
- No easy way to study the global convergence of the algorithm.
- The number of hyper-hyperparameters $(\alpha_{\epsilon_0}, \beta_{\epsilon_0})$, $\left(\alpha_{\xi_z}, \beta_{\xi_z}\right)$, $\left(\alpha_{\zeta_0}, \beta_{\zeta_0}\right)$, $(\alpha_{z_0}, \beta_{z_0})$ to be fixed is important. However, the results are not so sensitive to these parameters.

8.4 Gauss-Markov-Potts models

To introduce the Gauss-Markov-Potts model, let us have a look at the images in **Figure 11**.

The question we want to answer is: which prior model can be more appropriate for these images? One way, to answer to this question is either look at the histogram of the pixels or the pixels of the gradient images. Then, if we take one typical line of the gradient or one typical line of the image itself and draw them as a 1D-signal, we obtain the cases in **Figure 12**.

From these two figures, we see that for some cases, a Gaussian or generalized Gaussian may be very good models. But, for other cases, if we want to explicitly account for the presence of the contours, we can introduce a binary hidden variable to represent it. Finally, for the last example of the image in **Figure 10** and its corresponding typical line in **Figure 11**, we need to introduce a hidden variable $\boldsymbol{z}$ which encodes the following fact that:

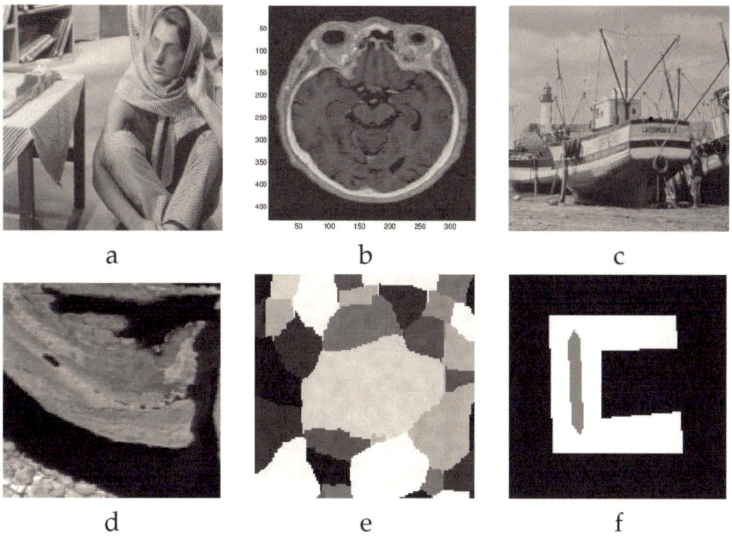

Figure 11.
Different images with different characteristics in different imaging systems.

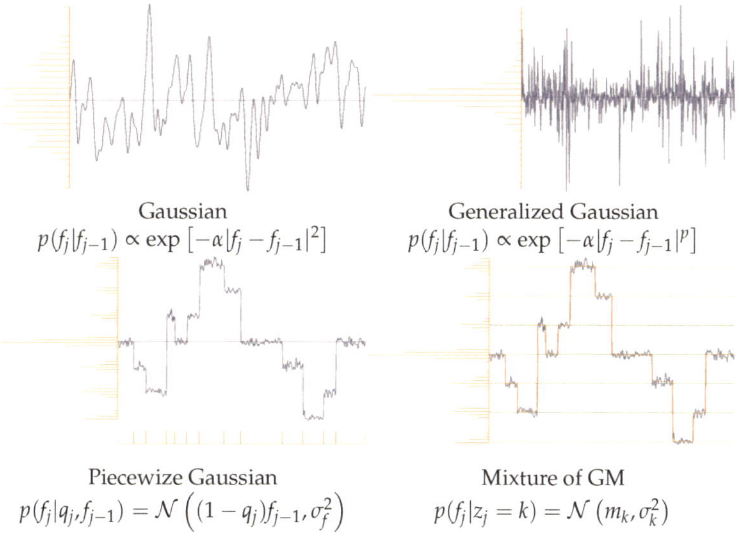

Figure 12.
Different possible prior modeling in relation to the different images of ***Figure 11****.*

In NDT applications of CT, the objects are, in general, composed of a finite number of materials, and the voxels corresponding to each material are grouped in compact regions.

How to model this prior information?

To answer to this question, first consider such an image $f(\mathbf{r})$ with its segmentation $z(\mathbf{r})$ and contours $q(\mathbf{r})$ as shown in **Figure 13**.

As it can be seen, we introduced two hidden variables $z(\mathbf{r})$ and $q(\mathbf{r})$, the first representing the segmentation and the second the contours of the image. $z(\mathbf{r})$ takes the integer values $\{k = 1, \cdots, K\}$, each presented by a different color and $q(\mathbf{r})$ a binary

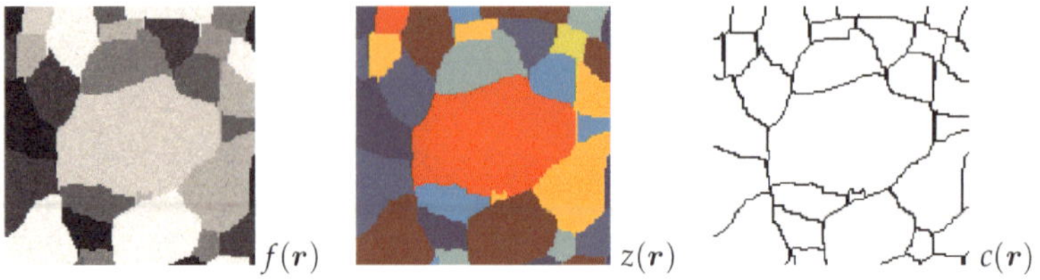

Figure 13.
An image of an object composed of homogeneous compact regions, its segmentation and the contours of thoses regions.

value $\{0, 1\}$. The second can easily be obtained from the first. So, from now, we consider only $z(\boldsymbol{r})$.

As each value of z represents a homogeneous material, we can translate this by:

$$p(f(\boldsymbol{r})|z(\boldsymbol{r}) = k, m_k, v_k) = \mathcal{N}(m_k, v_k) \tag{72}$$

encoding the fact that inside each homogeneous material, i.e.; all the pixels having $z(r) = k$, represent a homogenous material characterized by the two parameters f (m_k, v_k). This results to:

$$p(f(\boldsymbol{r})) = \sum_k P(z(\boldsymbol{r}) = k)\mathcal{N}(m_k, v_k) \text{ Mixture of Gaussians} \tag{73}$$

which shows the mixture of Gaussian model of the pixel values. See also **Figure 14**.

The next step is to propose a probability distribution for $\boldsymbol{z}$. As we want a compactness of the regions, a Markov modeling is appropriate:

$$p(z(\boldsymbol{r})|z(\boldsymbol{r}'), \boldsymbol{r}' \in \mathcal{V}(\boldsymbol{r})) \propto \exp\left[-\gamma \sum_{\boldsymbol{r}' \in \mathcal{V}(\boldsymbol{r})} \delta(z(\boldsymbol{r}) - z(\boldsymbol{r}'))\right] \tag{74}$$

A Potts Markov model is still more appropriate:

$$p(z(\boldsymbol{r}), \boldsymbol{r} \in \Omega) \propto \exp\left[-\gamma \sum_{\boldsymbol{r} \in \Omega} \sum_{\boldsymbol{r}' \in \mathcal{V}(\boldsymbol{r})} \delta(z(\boldsymbol{r}) - z(\boldsymbol{r}'))\right] \tag{75}$$

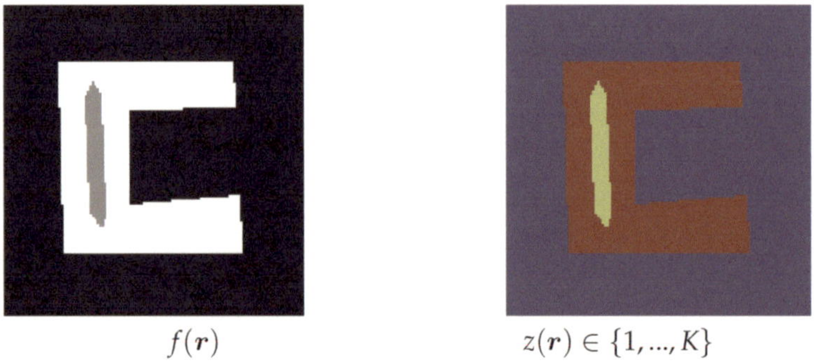

Figure 14.
A metalic object with a default area inside it: Black pixels represent air, white pixels metal and gray pixels the defaults area. On left image these are codes by colors ($z = 1$ represents air, $z = 2$ represents metal and $z = 3$ represents default area.

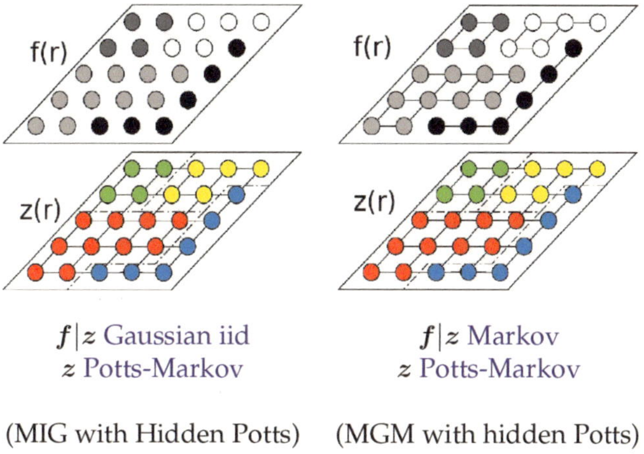

Figure 15.
Two proposed gauss-Markov-Potts models used in many NDT applications.

where Ω represents all pixels of the image.

Thus, to each pixel of the image is associated 2 variables $f(\boldsymbol{r})$ and $z(\boldsymbol{r})$ with the following possible properties:

- $\boldsymbol{f}|\boldsymbol{z}$ Gaussian iid, $\boldsymbol{z}$ iid: Mixture of Gaussians
- $\boldsymbol{f}|\boldsymbol{z}$ Gauss-Markov, $\boldsymbol{z}$ iid: Mixture of Gauss-Markov
- $\boldsymbol{f}|\boldsymbol{z}$ Gaussian iid, $\boldsymbol{z}$ Potts-Markov: Mixture of Independent Gaussians, (MIG with Hidden Potts)
- $\boldsymbol{f}|\boldsymbol{z}$ Markov, $\boldsymbol{z}$ Potts-Markov: Mixture of Gauss-Markov, (MGM with hidden Potts)

From these four different cases, we consider two which are illustrated in **Figure 15**.

Using the notations on this figure, and noting by $\boldsymbol{f}$ all the pixels of the image, by $\boldsymbol{z}$ all the pixels of the segmented image, and by $\boldsymbol{\theta}$ all the parameters $\{v_\epsilon, (\alpha_k, m_k, v_k), k = 1, \cdots, K\}$, we can write:

$$p(\boldsymbol{f}, \boldsymbol{z}, \boldsymbol{\theta}|\boldsymbol{g}) \propto p(\boldsymbol{g}|\boldsymbol{f}, \boldsymbol{v}_\epsilon) p(\boldsymbol{f}|\boldsymbol{z}, \boldsymbol{m}, \boldsymbol{v}) p(\boldsymbol{z}|\gamma, \alpha) p(\boldsymbol{\theta}) \quad (76)$$

where

$$\boldsymbol{m} = \{m_k, k = 1, \cdot, K\}, v = \{v_k, k = 1, \cdot, K\}, \alpha = \{\alpha_k, k = 1, \cdot, K\}, \boldsymbol{\theta} = \{v_\epsilon, m, v, alphab\}$$

The expressions of $p(\boldsymbol{g}|\boldsymbol{f}, v_\epsilon), p(\boldsymbol{f}|\boldsymbol{z}, \boldsymbol{m}, \boldsymbol{v})$ and $p(\boldsymbol{z}|\gamma, \boldsymbol{\alpha})$ have been given before. We need to define $p(\boldsymbol{\theta})$ which can be chosen as the conjugate priors: Dirichlet for α, Gaussian for $\boldsymbol{m}$ and Inverse-Gamma for all the variances.

Direct computation and use of $p(\boldsymbol{f}, \boldsymbol{z}, \boldsymbol{\theta}|\boldsymbol{g}; \mathcal{M})$ is too complex, because we do not have analytical expression for the proportionality term of the joint probability law:

$$p(\boldsymbol{f}, \boldsymbol{z}, \boldsymbol{\theta}|\boldsymbol{g}) \propto p(\boldsymbol{g}|\boldsymbol{f}, \boldsymbol{z}, \boldsymbol{\theta}) p(\boldsymbol{f}|\boldsymbol{z}, \boldsymbol{\theta}) p(\boldsymbol{z}) p(\boldsymbol{\theta}) \quad (77)$$

As we have three sets of variables $\boldsymbol{f}$, $\boldsymbol{z}$ and $\boldsymbol{\theta}$, we can use different schemes, for example a Gibbs sampling scheme:

$$\hat{\boldsymbol{f}} \sim p\left(\boldsymbol{f}|\hat{\boldsymbol{z}},\hat{\boldsymbol{\theta}},\boldsymbol{g}\right) \rightarrow \hat{\boldsymbol{z}} \sim p\left(\boldsymbol{z}|\hat{\boldsymbol{f}},\hat{\boldsymbol{\theta}},\boldsymbol{g}\right) \rightarrow \hat{\boldsymbol{\theta}} \sim \left(\boldsymbol{\theta}|\hat{\boldsymbol{f}},\hat{\boldsymbol{z}},\boldsymbol{g}\right) \tag{78}$$

with:

- Sample $\boldsymbol{f}$ from $p\left(\boldsymbol{f}|\hat{\boldsymbol{z}},\hat{\boldsymbol{\theta}},\boldsymbol{g}\right) \propto p(\boldsymbol{g}|\boldsymbol{f},\boldsymbol{\theta})p\left(\boldsymbol{f}|\hat{\boldsymbol{z}},\hat{\boldsymbol{\theta}}\right)$

  Needs optimisation of a quadratic criterion.

- Sample $\boldsymbol{z}$ from $p\left(\boldsymbol{z}|\hat{\boldsymbol{f}},\hat{\boldsymbol{\theta}},\boldsymbol{g}\right) \propto p\left(\boldsymbol{g}|\hat{\boldsymbol{f}},\hat{\boldsymbol{z}},\hat{\boldsymbol{\theta}}\right)p(\boldsymbol{z})$

  Needs sampling of a Potts Markov field.

- Sample $\boldsymbol{\theta}$ from $p\left(\boldsymbol{\theta}|\hat{\boldsymbol{f}},\hat{\boldsymbol{z}},\boldsymbol{g}\right) \propto p\left(\boldsymbol{g}|\hat{\boldsymbol{f}},\sigma_\epsilon^2 I\right)p\left(\hat{\boldsymbol{f}}|\hat{\boldsymbol{z}},(m_k,v_k)\right)p(\boldsymbol{\theta})$

More details and other schemes such as JMAP and VBA can be found in Refs. [26].

To illustrate an example of application, we considered a NDT application, where a metalic object is tested to detect a default inside it. As, the problem was, not only to detect the default, but also to characterize its shape and size, an X-ray computed tomography (CT) with only two projections is proposed and used. This problem is illustrated in **Figure 16**.

The mathematical part of this very ill-posed inverse problem is the following:

Given the functions $g_1(x)$ and $g_2(y)$ find the image $f(x,y)$.

This problem also arise in probability theory and statistics, where $f(x,y)$ is a joint distribution and $g_1(x)$ and $g_2(y)$ its two marginals. We know that this problem has infinite number of solutions: $f(x,y) = g_1(x)g_2(y)\Omega(x,y)$ where $\Omega(x,y)$ is called a Copula:

$$\int \Omega(x,y)\mathrm{d}x = 1 \text{ and } \int \Omega(x,y)\mathrm{d}y = 1$$

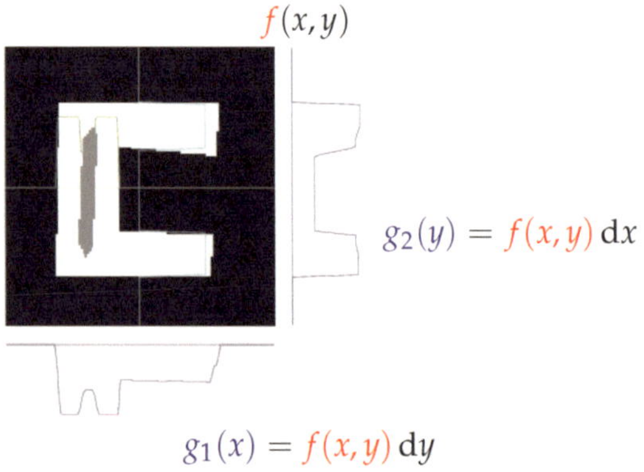

Figure 16.
A non destructive testing (NDT) application where $f(x,y)$ has to be reconstructed from its marginals $g_1(x)$ and $g_2(y)$.

So, any arbitrary copula function defines a solution. The problem is ill-posed and we need to use any possible prior information to try to obtain a unique or acceptable solution. The probabilistic solution we proposed is illustrated in **Figure 17**.

Unsupervised Bayesian estimation:

$$p(\boldsymbol{f}, \boldsymbol{z}, \boldsymbol{\theta} | \boldsymbol{g}) \propto p(\boldsymbol{g} | \boldsymbol{f}, \boldsymbol{z}, \boldsymbol{\theta}) p(\boldsymbol{f} | \boldsymbol{z}, \boldsymbol{\theta}) p(\boldsymbol{\theta})$$

A summary of the results is given in **Figure 18** where the proposed method result.

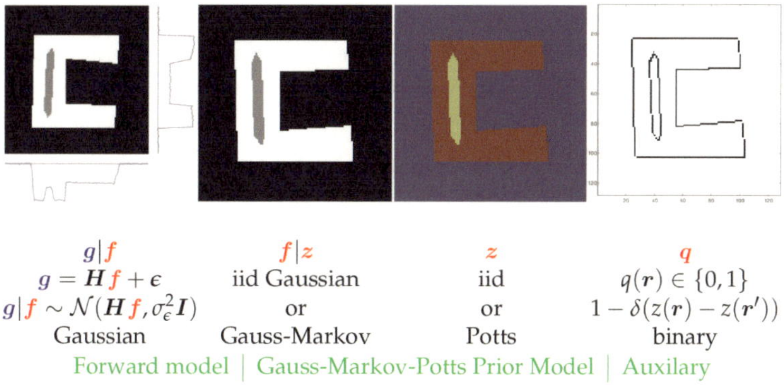

Figure 17.
Probabilistic Bayesian method for the NDT image resonstruction problem.

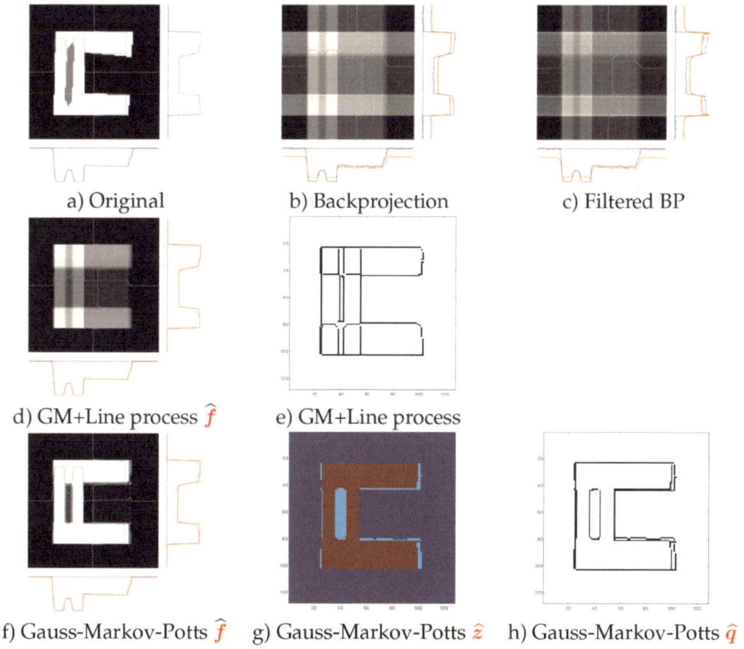

Figure 18.
Probabilistic Bayesian method for the NDT image resonstruction problem. a) Shows the original image f, b) is the result of Back-projection, c) is the result of filtered Back-projection, d) and e) are the result of a Markov model with hidden line process, and f), g) and h) show the results of the gauss-Markov-Potts method.

9. Bayesian computation

As we could see, very often, we can find the expression of the posterior law $p(\boldsymbol{f}|\boldsymbol{g})$, sometimes exactly as is the case of the linear models with Gaussian priors in the previous section, but often up to the normalization constant (the evidence term) $p(\boldsymbol{g})$ in:

$$p(\boldsymbol{f}|\boldsymbol{g}) = \frac{1}{p(\boldsymbol{g})} p(\boldsymbol{g}|\boldsymbol{f}) p(\boldsymbol{f}) = \frac{1}{p(\boldsymbol{g})} p(\boldsymbol{g}, \boldsymbol{f}). \tag{79}$$

This term is not necessary for Maximum A Posteriori (MAP) but it is needed for Expected A Posteriori (EAP) and for doing any other expectation computation.

This is the case, almost in all *Non-Gaussian prior models* or *Non-Gaussian noise models* or the *Non-Linear forward models*. In this chapter, a few cases are considered more in detail. Even in the Gaussian and linear case which is the simplest case, and we have analytical expressions for almost everything, the computational cost for large scale problems brings us to search for approximate but fast solutions.

9.1 Large scale linear and Gaussian models

As we could see in previous chapter, the linear forward model $\boldsymbol{g} = \boldsymbol{H}\boldsymbol{f} + \boldsymbol{\epsilon}$ with Gaussian noise and Gaussian prior is the simplest case where we can do all the computations analytically.

$$\left\{\begin{array}{l} p(\boldsymbol{g}|\boldsymbol{f}) = \mathcal{N}(\boldsymbol{g}|\boldsymbol{H}\boldsymbol{f}, v_\epsilon \boldsymbol{I}) \\ p(\boldsymbol{f}) = \mathcal{N}(\boldsymbol{f}|\boldsymbol{f}_0, v_f \boldsymbol{I}) \end{array}\right. \rightarrow \left\{\begin{array}{l} p(\boldsymbol{g}) = \mathcal{N}(\boldsymbol{g}|\boldsymbol{H}\boldsymbol{f}_0, v_f \boldsymbol{H}\boldsymbol{H}' + v_\epsilon \boldsymbol{I}), \\ p(\boldsymbol{f}|\boldsymbol{g}) = \mathcal{N}\left(\boldsymbol{f}|\hat{\boldsymbol{f}}, \hat{\Sigma}\right) \quad \text{with :} \\ \hat{\boldsymbol{f}} = \boldsymbol{f}_0 + [\boldsymbol{H}'\boldsymbol{H} + \lambda I]^{-1} \boldsymbol{H}'(\boldsymbol{g} - \boldsymbol{H}\boldsymbol{f}_0) \\ \hat{\Sigma} = v_\epsilon [\boldsymbol{H}'\boldsymbol{H} + \lambda \boldsymbol{I}]^{-1} \lambda = \frac{v_\epsilon}{v_f} \end{array}\right. \tag{80}$$

The trick here is, for example, for computing $\hat{\boldsymbol{f}}$ to use the fact that

$$\left\{\begin{array}{l} p(\boldsymbol{g}|\boldsymbol{f}) \propto \exp\left[-\frac{1}{2v_\epsilon}\|\boldsymbol{g} - \boldsymbol{H}\boldsymbol{f}\|_2^2\right] \\ p(\boldsymbol{f}) \propto \exp\left[-\frac{1}{2v_f}\|\boldsymbol{f}\|_2^2\right] \\ p(\boldsymbol{f}|\boldsymbol{g}) \propto \exp\left[-\frac{1}{2v_\epsilon} J(\boldsymbol{f})\right] \text{ with } J(\boldsymbol{f}) = \frac{1}{2}\|\boldsymbol{g} - \boldsymbol{H}\boldsymbol{f}\|^2 + \lambda\|\boldsymbol{f}\|_2^2, \quad \lambda = \frac{v_\epsilon}{v_f} \end{array}\right. \tag{81}$$

and, as the mean and the mode of a Gaussian probability law are the same, we can use:

$$\hat{\boldsymbol{f}} = \underset{\boldsymbol{f}}{\operatorname{argmax}}\{J(\boldsymbol{f})\} \text{ with } J(\boldsymbol{f}) = \|\boldsymbol{g} - \boldsymbol{H}\boldsymbol{f}\|^2 + \lambda\|\boldsymbol{f}\|^2 \tag{82}$$

and so the problem can be cast as an *optimization* problem of a quadratic criterion for which there are a great number of algorithms. Let here to show the simplest one which is the gradient based and so needs the expression of the gradient:

$$\nabla J(\boldsymbol{f}) = -2\boldsymbol{H}'(\boldsymbol{g} - \boldsymbol{H}\boldsymbol{f}) + 2\lambda\boldsymbol{f} \tag{83}$$

which can be summarized as follows:

$$\begin{cases} \boldsymbol{f}^{(0)} = 0 \\ \boldsymbol{f}^{(k+1)} = \boldsymbol{f}^{(k)} + \alpha\left[\boldsymbol{H}'\left(\boldsymbol{g} - \boldsymbol{H}\boldsymbol{f}^{(k)}\right) + 2\lambda\boldsymbol{f}^{(k)}\right] \end{cases} \tag{84}$$

As we can see, at each iteration, we need to be able to compute the *forward operation* $\boldsymbol{H}\boldsymbol{f}$ and the *backward operation* $\boldsymbol{H}'\delta\boldsymbol{g}$ where $\delta g = \boldsymbol{g}–\boldsymbol{H}\boldsymbol{f}$. This optimization algorithm needs to write two programs:

- Forward operation $\boldsymbol{H}\boldsymbol{f}$
- Adjoint operation $\boldsymbol{H}'\delta\boldsymbol{g}$

These two operations can be implemented using High Performance parallel processors such as Graphical Processor Units (GPU).

The computation of the posterior covariance is much more difficult. There are a few methods: The first category is the methods which use the particular structure of the matrix $\boldsymbol{H}$ and $\boldsymbol{H}'\boldsymbol{H}$ or $\boldsymbol{H}\boldsymbol{H}'$ as we can use the matrix inversion lemma and see that

$$\hat{\Sigma} = v_\epsilon[\boldsymbol{H}'\boldsymbol{H} + \lambda\boldsymbol{I}]^{-1} = v_\epsilon\boldsymbol{I} - \boldsymbol{H}'\left[\boldsymbol{H}\boldsymbol{H}' + \lambda^{-1}\boldsymbol{I}\right]^{-1} \tag{85}$$

For example, in a signal deconvolution problem, the matrix $\boldsymbol{H}$ has a Toeplitz structure and so have the matrices $\boldsymbol{H}'\boldsymbol{H}$ and $\boldsymbol{H}\boldsymbol{H}'$ which can be approximated by Circulant matrices and be diagonalized using the Fourier Transform.

The second, more general, is to approximate $\hat{\Sigma}$ by a diagonal matrix, which can also be interpreted as to approximate the posterior law $p(\boldsymbol{f}|\boldsymbol{g})$ by a separable $q(\boldsymbol{f}) = \Pi_j q\left(f_j\right)$. This brings us naturally to the Approximate Bayesian Computation (ABC). But, before going to the details of ABC methods, let consider the case where the hyperparameters of the problem (parameters of the prior laws) are also unknown.

To be able to do the computation, we need mainly to compute the determinant of the matrix $v_f\boldsymbol{H}\boldsymbol{H}' + v_\epsilon\boldsymbol{I}$ for $p(\boldsymbol{g})$ and the inverse of the matrices $[\boldsymbol{H}'\boldsymbol{H} + \lambda\boldsymbol{I}]$ or $\left[\boldsymbol{H}\boldsymbol{H}' + \lambda^{-1}\boldsymbol{I}\right]$.

9.2 Large scale computation of the posterior covariance

Computing the determinant of the matrix $v_f\boldsymbol{H}\boldsymbol{H}' + v_\epsilon\boldsymbol{I}$ for $p(\boldsymbol{g})$ and the inverse of the matrices $[\boldsymbol{H}'\boldsymbol{H} + \lambda\boldsymbol{I}]$ or, $\left[\boldsymbol{H}\boldsymbol{H}' + \lambda^{-1}\boldsymbol{I}\right]$ which are needed for uncertainty quantification, are between the greatest subjects of open research for *Big Data* problems. Here, we consider a few cases.

9.2.1 Structured matrices

One solution is to use the particular structure of these matrices when possible. This is the case for deconvolution or image restoration, where these matrices have Toeplitz or Bloc-Toeplitz structures which can be well approximated by Circulant or Bloc-Circulant matrices and diagonalized using Fourier Transform (FT) and Fast FT

(FFT). The main idea here is using the properties of the circulant matrices: If $\boldsymbol{H}$ is a circulant matrix, then

$$\boldsymbol{H} = \boldsymbol{F}\Lambda\boldsymbol{F}' \tag{86}$$

where $\boldsymbol{F}$ is the DFT or FFT matrix and $\boldsymbol{F}'$ the IDFT or IFFT and Λ is a diagonal matrix whose elements are the FT of the first line of the circulent matrix. As the first line of that circulent matrix contains the samples of the impulse response, the vector of the diagonal elements represents the spectrum of the impulse response (transfer function). Using this property, we have:

$$[\boldsymbol{H}'\boldsymbol{H} + \lambda\boldsymbol{I}]^{-1} = [\boldsymbol{F}'\Lambda\boldsymbol{F}\boldsymbol{F}'\Lambda + \lambda]^{-1} = [\boldsymbol{F}'\Lambda^2\boldsymbol{F} + \lambda\boldsymbol{I}]^{-1} = \boldsymbol{F}[\Lambda^2 + \lambda\boldsymbol{I}]^{-1}\boldsymbol{F}' \tag{87}$$

9.2.2 Sampling based methods

Second solution is generating samples from the posterior law and use them to compute the variances and covariances. So, the problem is how to generate a sample from the posterior law

$$\begin{cases} p(\boldsymbol{f}|\boldsymbol{g}) = \mathcal{N}\left(\boldsymbol{f}|\hat{\boldsymbol{f}}, \hat{\Sigma}\right) \quad \text{with :} \\ \hat{\boldsymbol{f}} = \boldsymbol{f}_0 + [\boldsymbol{H}'\boldsymbol{H} + \lambda\boldsymbol{I}]^{-1}\boldsymbol{H}'(\boldsymbol{g} - \boldsymbol{H}\boldsymbol{f}_0) \\ \hat{\Sigma} = v_\epsilon[\boldsymbol{H}'\boldsymbol{H} + \lambda\boldsymbol{I}]^{-1}, \quad \lambda = \dfrac{v_\epsilon}{v_f} \end{cases} \tag{88}$$

One solution is to compute the Cholesky decomposition of the covariance matrix $\hat{\Sigma} = \boldsymbol{A}\boldsymbol{A}'$, generate a vector, $\boldsymbol{u} \sim \mathcal{N}(\boldsymbol{u}|0, \boldsymbol{I})$ and then generate a sample $\boldsymbol{f} = \boldsymbol{A}\boldsymbol{u} + \hat{\boldsymbol{f}}$ [27]. We can compute $\hat{\boldsymbol{f}}$ by optimizing

$$J(\boldsymbol{f}) = \frac{1}{2}\|\boldsymbol{g} - \boldsymbol{H}\boldsymbol{f}\|^2 + \lambda\|\boldsymbol{f} - \boldsymbol{f}_0\|_2^2, \lambda = \frac{v_\epsilon}{v_f}, \tag{89}$$

but the main computational cost is the Cholesky factorization.

Another approach, called Perturbation-Optimization [28, 29] is based on the following property:

If we note $\boldsymbol{x} = \boldsymbol{f} + [\boldsymbol{H}'\boldsymbol{H} + \lambda\boldsymbol{I}]^{-1}\boldsymbol{H}'(\boldsymbol{g} - \boldsymbol{H}\boldsymbol{f})$ and look for its expected and covariance matrix, it can be shown that:

$$\begin{cases} \mathrm{E}\{\boldsymbol{x}\} = \hat{\boldsymbol{f}} \\ \mathrm{Cov}\,[\boldsymbol{x}] = \hat{\Sigma} \end{cases} \tag{90}$$

So, to generate a sample from the posterior law, we can do the following:

- Generate two random vectors $\boldsymbol{\epsilon}_f \sim \mathcal{N}(\boldsymbol{\epsilon}_f|0, v_f I)$ and $\boldsymbol{\epsilon}_g \sim \mathcal{N}(\boldsymbol{\epsilon}_g|0, v_\epsilon I)$;

- Define $\tilde{\boldsymbol{g}} = \boldsymbol{g} + \boldsymbol{\epsilon}_g$ and $\tilde{\boldsymbol{f}} = \boldsymbol{f} + \boldsymbol{\epsilon}_f$ and optimize

$$J(\tilde{\boldsymbol{f}}) = \frac{1}{2}\|\tilde{\boldsymbol{g}} - \boldsymbol{H}\boldsymbol{f}\|^2 + \lambda\|\tilde{\boldsymbol{f}} - \boldsymbol{f}_0\|_2^2 \tag{91}$$

- The obtained solution $\boldsymbol{f}^{(n)} = \arg\min_{\tilde{f}}\left\{J\left(\tilde{f}\right)\right\}$ is a sample from the desired posterior law.

By repeating this process for a great number of times, we can use them to obtain good approximations for the posterior mean $\hat{\boldsymbol{f}}$ and the posterior covariance $\hat{\Sigma}$ by computing their empirical mean values. We need however fast and accurate optimization algorithms.

10. References to examples of applications

The above mentioned methods have been used with success in different applications:

- Medical imaging and Computed tomography (CT) [30–36].
- Diffraction tomography and Microwave imaging [37–42].
- 3D Computed Tomography [18, 22, 43]
- Acoustical imaging [44–49]
- Hyperspectral imaging [50]
- Spectrometry [51]
- Eddy current tomography [52]
- Non destructive testing applications [53]
- Emission Tomography [54]
- SAR imaging [55]
- Chronobiological time series [56]

11. Conclusions

Mainly, in this chapter, first we described inverse problems and gave a few classical examples such as deconvolution, image restoration, computed tomography X-ray image reconstruction, Fourier synthesis inversion problem which arise in many imaging systems. Then, we mentioned that there are two classes of methods for inverse problems: deterministic regularization and Bayesian inference methods. Then, we started by describing the Bayesian parameter estimation. The main parts of the chapter is focused on Bayesian inference for inverse problems. We saw that the main difficulty is the great dimension of unknown quantities and the appropriate choice of the prior law. For this, first we described many simple and hierarchical prior models which are used in real applications. For the second main difficulty, which is the computational aspects, we described different approximate Bayesian computations

(ABC) and in particular the variational Bayesian approximation (VBA) methods and showed how to use these methods, for example for hyperparameter estimation or for large scale inverse problems.

Author details

Ali Mohammad-Djafari[1,2]

1 CNRS, France

2 ISCT, Bures-sur-Yvette, France

*Address all correspondence to: djafari@ieee.org

References

[1] Idier J. Approche bay´esienne pour les probl`emes inverses. Herm`es Science Publications; 2001

[2] Mohammad-Djafari A. Inverse Problems in Vision and 3D Tomography. ISTE-WILEY; 2010

[3] Mohammad-Djafari A. Efficient scalable variational bayesian approximation methods for inverse problems. In: SIAM Uncertainty Quantification UQ16. EPFL; April 2016

[4] Idier J. Bayesian Approach to Inverse Problems. John Wiley & Sons; 2008

[5] Mohammad-Djafari A. Probl`emes inverses en imagerie et en vision en deux volumes ins´eparables. In: Trait´e Signal et Image, IC2. ISTE-WILEY; 2009

[6] Carasso AS. Direct blind deconvolution. SIAM Journal on Applied Mathematics. 2001;**61**(6):1980-2007

[7] Chan T, Wong C-K. Total variation blind deconvolution. IEEE Transactions on Image Processing. 1998;**7**(3):370-375

[8] Kak AC, Slaney M. Principles of Computerized Tomographic Imaging. SIAM; 2001

[9] Jackson JI, Meyer CH, Nishimura DG, Macovski A. Selection of a convolution function for Fourier inversion using gridding [computerized tomography application]. IEEE Transactions on Medical Imaging. 1991;**10**(3):473-478

[10] Chapdelaine C, Mohammad-Djafari A, Gac N, Parra E. A 3D Bayesian computed tomography reconstruction algorithm with Gauss-Markov-Potts prior model and its application to real data. Fundamenta Informaticae. [submitted]

[11] Osher S, Burger M, Goldfarb D, Xu J, Yin W. An iterative regularization method for total variation-based image restoration. Multiscale Modeling & Simulation. 2005;**4**(2):460-489

[12] Wang Y, Yang J, Yin W, Zhang Y. A new alternating minimization algorithm for total variation image reconstruction. SIAM Journal on Imaging Sciences. 2008;**1**(3):248-272

[13] Goldstein T, Osher S. The split Bregman method for L1-regularized problems. SIAM Journal on Imaging Sciences. 2009;**2**(2):323-343

[14] Bertocchi C, Chouzenoux E, Corbineau M-C, Pesquet J-C, Prato M. Deep unfolding of a proximal interior point method for image restoration. Inverse Problems. 2019;**36**

[15] Mohammad-Djafari A. Gauss-Markov-Potts priors for images in computer tomography resulting to joint optimal reconstruction and segmentation. International Journal of Tomography and Statistics (IJTS). 2008; **11**:76-92

[16] Ayasso H, Mohammad-Djafari A. Joint NDT image restoration and segmentation using Gauss–Markov–Potts prior models and variational Bayesian computation. IEEE Transactions on Image Processing. 2010; **19**(9):2265-2277

[17] Feron O, Duchene B, Mohammad-Djafari A. Microwave imaging of inhomogeneous objects made of a finite number of dielectric and conductive materials from experimental data. Inverse Problems. 2005;**21**(6):S95

[18] Wang L, Gac N, Mohammad-Djafari A. Bayesian 3D X-ray computed

tomography image reconstruction with a scaled Gaussian mixture prior model. AIP Conference Proceedings. 2015;**1641**: 556-563

[19] Chapdelaine C, Mohammad-Djafari A, Gac N, Parra E. A joint segmentation and reconstruction algorithm for 3D Bayesian computed tomography using Gauss-Markov-Potts prior model. In: The 42nd IEEE International Conference on Acoustics, Speech and Signal Processing (ICASSP). 2017

[20] Chapdelaine C, Gac N, Mohammad-Djafari A, Parra E. New GPU implementation of separable footprint projector and backprojector : First results. In: The 5th International Conference on Image Formation in X-Ray Computed Tomography. 2018

[21] Chapdelaine C. Variational Bayesian approach and Gauss-Markov-Potts prior model. arXiv:1808.09552. 2018

[22] Wang L, Mohammad-Djafari A, Gac N. X-ray computed tomography using a sparsity enforcing prior model based on haar transformation in a Bayesian framework. Fundamenta Informaticae. 2017;**155**(4):449-480

[23] Bioucas-Dias JM, Figueiredo MAT. An iterative algorithm for linear inverse problems with compound regularizers. In: 15th IEEE International Conference on Image Processing, 2008 (ICIP 2008). IEEE; 2008. pp. 685-688

[24] Florea MI, Vorobyov SA. A robust fista-like algorithm. In: 2017 IEEE International Conference on Acoustics, Speech and Signal Processing (ICASSP). IEEE; 2017. pp. 4521-4525

[25] Yu S, Wu Z, Xu X, Wohlberg B, Kamilov US. Scalable plug-and-play ADMM with convergence guarantees. IEEE Transactions on Computational Imaging. 2021;7:849-863

[26] Chapdelaine C. Reconstruction 3D par rayons X pour le Contrˆole Non Destructif de pi`eces a´eronautique [Th`ese]. Orsay, France: Universit´e de Paris–Sud; 2019

[27] Gilavert C, Moussaoui S, Idier J. Efficient Gaussian sampling for solving large-scale inverse problems using MCMC. IEEE Transactions on Signal Processing. 2015;**63**(1):70-80

[28] Giovannelli J-F. Estimation of the Ising field parameter thanks to the exact partition function. In: ICIP. 2010. pp. 1441-1444

[29] Orieux F, Feron O, Giovannelli J-F. Sampling high dimensional Gaussian distributions for general linear inverse problems. IEEE Signal Processing Letters. 2012;**19**(5):251-254

[30] Mohammad-Djafari A, Demoment G. Maximum entropy image reconstruction in X-ray and diffraction tomography. IEEE Transactions on Medical Imaging. 1988;**7**(4):345-354

[31] Soussen C, Mohammad-Djafari A. Polygonal and polyhedral contour reconstruction in computed tomography. IEEE Transactions on Image Processing. 2004;**13**(11):1507-1523

[32] Wang L, Mohammad-Djafari A, Gac N. Bayesian method with sparsity enforcing prior of dual-tree complex wavelet transform coefficients for X-ray CT image reconstruction. In: 2017 25th European Signal Processing Conference (EUSIPCO). 2017. pp. 478-482

[33] Mohammad-Djafari A. Hierarchical markov modeling for fusion of X ray radiographic data and anatomical data in

computed tomography. In: Proceedings IEEE International Symposium on Biomedical Imaging. July 2002. pp. 401-404

[34] Mohammad-Djafari A, Sauer K, Khagu Y, Cano E. Reconstruction of the shape of a compact object from few projections. In: Proceedings of International Conference on Image Processing. Vol. 1. Oct 1997. pp. 165-168

[35] Wang L, Mohammad-Djafari A, Gac N. X-ray computed tomography simultaneous image reconstruction and contour detection using a hierarchical markovian model. In: 2017 IEEE International Conference on Acoustics, Speech and Signal Processing (ICASSP). March 2017. pp. 6070-6074

[36] Wang L, Mohammad-Djafari A, Gac N, Dumitru M. Computed tomography reconstruction based on a hierarchical model and variational Bayesian method. In: 2016 IEEE International Conference on Acoustics, Speech and Signal Processing (ICASSP). IEEE; March 2016. pp. 883-887

[37] Carfaatan H, Mohammad-Djafari A, Idier J. A single site update algorithm for nonlinear diffraction tomography. In: 1997 IEEE International Conference on Acoustics, Speech, and Signal Processing. Vol. 4. Apr 1997. pp. 2837-2840

[38] Nguyen MK, Mohammad-Djafari A. Bayesian approach with the maximum entropy principle in image reconstruction from microwave scattered field data. IEEE Transactions on Medical Imaging. 1994;**13**(2):254-262

[39] Gharsalli L, Duch^ene B, Mohammad-Djafari A, Ayasso H. Microwave tomography for breast cancer detection within a variational Bayesian approach. In: 21st European Signal Processing Conference (EUSIPCO 2013). Sept 2013. pp. 1-5

[40] Gharsalli L, Duch^ene B, Mohammad-Djafari A, Ayasso H. A gauss markov mixture prior model for a variational bayesian approach to microwave breast imaging. In: 2014 IEEE Conference on Antenna Measurements Applications (CAMA). Nov 2014. pp. 1-4

[41] Gharsalli L, Duch^ene B, Mohammad-Djafari A, Ayasso H. A gradient-like variational Bayesian approach: Application to microwave imaging for breast tumor detection. In: 2014 IEEE International Conference on Image Processing (ICIP). Oct 2014. pp. 1708-1712

[42] Ayasso H, Duch^ene B, Mohammad-Djafari A. A variational Bayesian approach for frequency diverse non-linear microwave imaging. In: 2012 19th IEEE International Conference on Image Processing. Sept 2012. pp. 2069-2072

[43] Gac N, Vabre A, Mohammad-Djafari A, Rabanal A, Buyens F. GPU implementation of a 3D Bayesian CT algorithm and its application on real foam reconstruction. In: The First International Conference on Image Formation in X-Ray Computed Tomography. 2010. pp. 151-155

[44] Chu N, Mohammad-Djafari A, Picheral J. A Bayesian sparse inference approach in near-field wideband aeroacoustic imaging. In: 2012 19th IEEE International Conference on Image Processing. Sept 2012. pp. 2529-2532

[45] Chu N, Mohammad-Djafari A, Gac N, Picheral J. A variational Bayesian approximation approach via a sparsity enforcing prior in acoustic imaging. In: 2014 13th Workshop on Information Optics (WIO). July 2014. pp. 1-4

[46] Chu N, Picheral J, Mohammad-Djafari A. A robust super-resolution approach with sparsity constraint for near-field wideband acoustic imaging. In: 2011 IEEE International Symposium on Signal Processing and Information Technology (ISSPIT). Dec 2011. pp. 310-315

[47] Chu N, Picheral J, Mohammad-Djafari A, Gac N. A robust super-resolution approach with sparsity constraint in acoustic imaging. Applied Acoustics. 2014;**76**(1):197-208

[48] Chu N, Zhao H, Yu L, Huang Q, Ning Y. Fast and high-resolution acoustic beamforming: A convolution accelerated deconvolution implementation. IEEE Transactions on Instrumentation and Measurement. 2020;**99**:1

[49] Chu N, Mohammad-Djafari A, Picheral J. Robust Bayesian superresolution approach via sparsity enforcing a priori for near-field aeroacoustic source imaging. Journal of Sound & Vibration. 2013;**332**(18):4369-4389

[50] Bali N, Mohammad-Djafari A. Bayesian approach with hidden markov modeling and mean field approximation for hyperspectral data analysis. IEEE Transactions on Image Processing. 2008; **17**(2):217-225

[51] Perenon R, Sage E, Mohammad-Djafari A, Duraffourg L, Hentz S, Brenac A, et al. Bayesian inversion of multi-mode NEMS mass spectrometry signal. In: 21st European Signal Processing Conference (EUSIPCO 2013). Sept 2013. pp. 1-5

[52] Premel D, Mohammad-Djafari A. Eddy current tomography in cylindrical geometry. IEEE Transactions on Magnetics. 1995;**31**(3):2000-2003

[53] Mohammad-Djafari A, Robillard L. Hierarchical markovian models for 3D computed tomography in non destructive testing applications. In: 2006 14th European Signal Processing Conference. Sept 2006. pp. 1-5

[54] Fall MD, Barat E, Comtat C, Dautremer T, Montagu T, Mohammad-Djafari A. A discrete-continuous bayesian model for emission tomography. In: 2011 18th IEEE International Conference on Image Processing. Sept 2011. pp. 1373-1376

[55] Zhu S, You P, Wang H, Li X, Mohammad-Djafari A. Recognition-oriented Bayesian sar imaging. In: 2011 3rd International Asia-Pacific Conference on Synthetic Aperture Radar (APSAR). Sept 2011. pp. 1-4

[56] Dumitru M, Mohammad-Djafari A. Periodic components estimation in chronobiological time series via a bayesian approach. In: 2015 23rd European Signal Processing Conference (EUSIPCO). Aug 2015. pp. 2246-2250

Chapter 3

Robust Bayesian Estimation

Ahmed Saadoon Mannaa

Abstract

Bayes methods in statistical inference are one of the important methods, and most of the research and messages tend to use the Bayes method in the estimation process. The regular Bayes method does not meet this problem, so in this thesis it is possible to verify the existence of prior data conflict by modeling the parameters of the prior distribution and then comparing the standard deviation of the prior distribution with the standard deviation of the posterior distribution, if the value of the standard deviation of the prior distribution is greater than the deviation. The standard distribution for the posterior distribution, it means that there is a problem of prior data conflict. Then we used an approach to solve this problem through a set of prior distributions called this approach by the robust Bayesian method, to identify the behavior of the estimators, two types of failure models were used, the first Weibull distribution to match it with continuous data. The second is a (Binomial) distribution to match the discrete data, the regular Bayes method is compared with the robust Bayesian method by using integrated mean square error (IMSE). In the Weibull distribution, the scale parameter (θ) and the survival function were estimated for two simulation experiments, the first was in the case of prior data unconflict the second was in the case of prior data conflict, so the simulation results showed that the robust Bayes method is the best by using the comparison criterion integrated mean square error (IMSE). On the practical side, real data were collected from Al-Manathira Hospital of the Najaf Health Department for the deaths of heart attack patients for 2018, the time of admission of the patient to the hospital until death was recorded, which is the time Exit where a sample of (15) patients was collected and the test of goodness of fit showed that the data follow a Weibull distribution with two parameters, the robust Bayes method was used to estimate the scale parameter and the survival function. As for the Binomial distribution, the parameter (P) and survival function were estimated for two experiments from the first simulation, which was in the case of prior data unconflict, as for the second experiment, it was in the case of prior data conflict. The simulation results showed that the robust Bayes method is the best by using the comparison criterion (IMSE). On the practical side, real data were collected from Yarmouk Teaching Hospital on breast cancer patients' mortality from 2010 to 2017, and the test of goodness of fit showed that the data follow a Binomial distribution, the robust Bayes method was used to estimate the parameter (P) and survival function.

Keywords: robust Bayesian, prior data conflict, survival function, iLuck model, regular Bayesian, Weibull distribution, binomial distribution

1. Introduction

Both the reliability function and the survival function have the same property, which is the measurement of the life span of a particular system or organism. In systems and equipment, it is called the reliability function, but for the organism, it is called the survival function.

Sometimes, especially in the analysis of survival functions, the failure events are few, so we need to include prior information, which can be used by the Bayes method. When merging the prior information with the observations to obtain the posterior distribution according to the Bayes rule, a problem may appear to us, which is the problem of prior data conflict with views, in regular Bayes method, this problem is not checked and is not addressed and thus unreal estimators are obtained, In this thesis, we will use an approach to address this problem, which is the prior data conflict problem, and thus this method is called the Robust Bayes Procedure.

What is meant by the prior data conflict is when the information of the prior distribution is combined with the distribution of observations, which may cause us this problem, meaning that the data under study are less homogeneous when the information of the prior distribution is combined with it, and thus we obtain unreal estimations without realizing.

In analyzing the problem, the researcher relied on two models of failure, the first model is the Weibull distribution with two parameters to match the continuous data, the second model is the Binomial distribution to match the discrete data to identify the behavior of the capabilities in these two types of data and the appropriateness of the robust methods to deal with the existence of the problem prior data conflict.

2. Robust Bayesian procedure

We also noted earlier that the concept of Bayes theory depends on prior information so that prior information is combined with the distribution of observations according to the Bayes rule, for the purpose of obtaining the posterior distribution, from here we may have a problem, which is a problem that prior data conflict. Whereas prior data they are the default values that are assumed for the parameters of the prior distribution, to find out this problem by updating the parameters of the prior distribution through two methods, namely Expected Conditional or Canonical Exponential Family and provided that the prior distribution is conjugate prior, after obtaining the prior distribution with the updated parameters, we extract the posterior distribution and then extract the standard deviation of the distribution if the value of the standard deviation of the prior distribution is greater than the standard deviation of the posterior distribution, then this means that there is a problem of prior data conflict, Thus, this problem can be addressed through the steps that we will explain later, This method is called the robust Bayesian method [1].

After the default values for the parameters of the prior distribution are chosen, the standard deviation of the prior distribution and the posterior distribution are extracted. If the value of the standard deviation of the prior distribution is greater than the standard deviation of the posterior distribution, this means that there is a problem of prior data conflict and provided that the posterior distribution is conjugate prior, this is the method that will be used in this chapter to verify the prior data conflict.

There are other ways to verify the prior data conflict that we did not used in this chapter, for example (Conflict checks based on relative belief, Connections between the relative belief and score checks and Other approaches to prior-data conflict checking) [2].

Then we move on to addressing the problem of prior data conflict through the proposal presented by (Walter and Augustin; 2009), this is for the purpose of generating a set of prior parameters, in short $\left(\prod^0 = [\underline{n}^0, \bar{n}^0]x\left[\underline{y}^0, \bar{y}^0\right]\right)$, So that this model that generates a set of prior parameters is called generalized iLuck-model and therefore we will get a set of posterior distributions, And then a Bayes estimator is obtained according to the type of loss function used, and thus this method is called the robust Bayesian method [3].

3. Weibull distribution

The Weibull distribution was used in 1951 by researcher Waldi Weibull in many experiments related to the reliability in the mechanical aspect and survival in the human aspect.

The emergence of this distribution, especially in the Second World War, and its wide applications in the field of reliability and life tests, was the focus of the attention of a number of researchers in this field, great in theory and practice.

The probability density function (pdf) for a two-parameter Weibull distribution is in the following form [4]:

$$f(t) = \frac{\beta}{\theta} t^{\beta-1} e^{-\frac{t^\beta}{\theta}}; \theta > 0, t > 0 \tag{1}$$

Since:

β: shape parameter.

θ: Scale parameter.

The formula for the (CDF) cumulative function is:

$$F(t) = 1 - e^{-\frac{t^\beta}{\theta}} \tag{2}$$

The formula for the survival function is:

$$S(t) = e^{-\frac{t^\beta}{\theta}} \tag{3}$$

and the formula for moment r^{th} is:

$$M_r = \theta^{\frac{r}{\beta}} \Gamma\left(1 + \frac{r}{\beta}\right) \tag{4}$$

3.1 Bayesian estimation of scale parameter for Weibull distribution

Suppose we have a sample that follows the Weibull distribution shown in Eq. (1) and the appropriate prior distribution for the parameter (θ) is inverse gamma distribution according to the following formula [5]:

$$f(\theta/a,b) = \frac{b^a}{\Gamma(a)}\theta^{-a-1}e^{-\frac{b}{\theta}} \tag{5}$$

By using the Bayes rule, we get the posterior distribution, as shown below [6]:

$$f(\theta/t) = \frac{f(t_1, t_2, \dots, t_n/\theta, \beta)f(\theta/a,b)}{\int_0^\infty f(t_1, t_2, \dots, t_n/\theta, \beta)f(\theta/a,b)d\theta} \tag{6}$$

$$= \frac{\left(\sum_{i=1}^n t_i^\beta + b\right)^{(a+n)}}{\Gamma(a+n)}\theta^{-(a+n)-1}e^{-\frac{\left(\sum_{i=1}^n t_i^\beta + b\right)}{\theta}}$$

Since:

$$\sum_{i=1}^n t_i^\beta = \tau(t)$$

$$f(\theta/t) \sim \text{Inverse Gamma}(a+n, \tau(t)+b)$$

Eq. (6) represents the posterior distribution for the parameter (θ), and according to the squared loss function, the Bayesian estimator for the parameter (θ) is the mean of the posterior distribution, as in the following steps [7]:

$$E(\theta/t) = \hat{\theta}$$

$$\hat{\theta} = \int_0^\infty \theta f(\theta/t)d\theta$$

$$= \int_0^\infty \theta \frac{(\tau(t)+b)^{(a+n)}}{\Gamma(a+n)}\theta^{-(a+n)-1}e^{-\frac{(\tau(t)+b)}{\theta}}d\theta$$

$$= \int_0^\infty \frac{(\tau(t)+b)^{(a+n)}}{\Gamma(a+n)}\theta^{-(a+n)-1+1}e^{-\frac{(\tau(t)+b)}{\theta}}d\theta$$

By using the transformation:

$$\text{Let } y = \frac{(\tau(t)+b)}{\theta} \Rightarrow \theta = \frac{(\tau(t)+b)}{y}, J = \frac{(\tau(t)+b)}{y^2}$$

$$\hat{\theta} = \int_0^\infty \frac{(\tau(t)+b)^{(a+n)}}{\Gamma(a+n)}\left(\frac{\tau(t)+b}{y}\right)^{-(a+n)} e^{-y}\frac{(\tau(t)+b)}{y^2}dy$$

$$\hat{\theta} = \frac{(\tau(t)+b)^{(a+n-a-n+1)}}{\Gamma(a+n)}\int_0^\infty \left(\frac{1}{y}\right)^{-(a+n)+2} e^{-y}dy$$

$$\hat{\theta} = \frac{(\tau(t) + b)}{\Gamma(a+n)} \int_0^\infty y^{a+n-2} e^{-y}\, dy$$

$$\hat{\theta} = \frac{(\tau(t) + b)}{\Gamma(a+n)} \Gamma(a+n-1)$$

$$\hat{\theta} = \frac{(\tau(t) + b)}{a+n-1} \tag{7}$$

3.2 Bayesian estimation of survival function for Weibull distribution

From Eq. (6), according to the squared loss function, the Bayesian estimator for the survival function is [8]:

$$\hat{S}(t) = \int_0^\infty S(t)\, f(\theta/t)\, d\theta$$

$$\hat{S}(t) = \left(\frac{b + \tau(t)}{b + \tau(t) + t_i^\beta}\right)^{a+n} \tag{8}$$

3.3 Checking for prior data conflict for Weibull distribution

Suppose we have a sample that follows a Weibull distribution according to the Eq. (1) and the prior distribution suitable for the scale parameter (θ) is Inverse Gamma because it is a conjugate prior and as in the Eq. (5) [9].

Then we need to update the parameters of the prior distribution so that it is $n^0 > 1, y^0 > 0$ instead of the parameters (a, b), through two methods we get the prior distribution as shown in the following steps [10]:

The first method: It is the Expected Conditional method, as shown in the following steps:

$$E(\theta/a, b) = y^0 = \frac{b}{a-1} = \frac{b}{n^0} \Rightarrow b = n^0 y^0, n^0 = a - 1 \Rightarrow a = n^0 + 1$$

Then the prior distribution with the updated parameters can be written in the following form:

$$f\left(\theta/n^0 y^0\right) \propto \frac{\left(n^0 y^0\right)^{n^0+1}}{\Gamma(n^0+1)} \theta^{-\left(n^0+1\right)-1} e^{-\frac{n^0 y^0}{\theta}} \tag{9}$$

Since:

y^0: Pre-guessing the scale parameter.

n^0: Pre- guessing the sample size.

The second method: In this method, the prior distribution can be determined with parameters $\left(n^0, y^0\right)$ through the following steps:

The first step: writing the model in the form of canonical exponential family and as shown below [10]:

$$f(x/\theta) = a(x) \exp\left(\psi.\tau(x) - nb(\psi)\right) \tag{10}$$

$$f(t/\theta,\beta) = \beta^n \prod_{i=1}^{n} t_i^{\beta-1} e^{-\sum_{i=1}^{n} t_i^{\beta}\theta^{-1} - n\ln(\theta)} \tag{11}$$

$$a(x) = \beta^n \prod_{i=1}^{n} t_i^{\beta-1}, \psi = -\frac{1}{\theta}, \tau(t) = \sum_{i=1}^{n} t_i^{\beta}, b(\psi) = \ln(\theta)$$

The second step: constructing the prior distribution with parameters (n^0, y^0) through the following form:

$$f(\psi/n^0, y^0)d\psi \propto \exp\{n^0[y^0.\psi - b(\psi)]\}d\psi \tag{12}$$

$$f(\psi/n^0, y^0)d\psi \propto \exp\left\{n^0\left[y^0\left(-\frac{1}{\theta}\right) - \ln(\theta)\right]\right\}d\psi$$

$$\left|\frac{d\psi}{d\theta}\right| = \frac{1}{\theta^2}$$

$$f(\theta/n^0, y^0)d\theta = f(\psi/n^0, y^0)\left|\frac{d\psi}{d\theta}\right|d\theta \propto \exp\left\{-\frac{n^0y^0}{\theta} - n^0\ln\theta\right\}\frac{1}{\theta^2}$$

$$f(\theta/n^0, y^0) \propto \theta^{-(n^0+1)-1} e^{-\frac{n^0y^0}{\theta}} \tag{13}$$

For the purpose of testing whether or not there is a problem of prior data conflict, we extract the standard deviation of the prior distribution and the standard deviation of the posterior distribution.

After we get the prior distribution with the updated parameters, we extract the standard deviation, as in the following steps:

$$M_r = \int_0^{\infty} \theta^r f(\theta/n^0, y^0)d\theta$$

$$M_r = \frac{(n^0y^0)^{n^0+1}}{\Gamma(n^0+1)} \int_0^{\infty} \theta^{-(n^0+1)-1+r} e^{-\frac{n^0y^0}{\theta}} d\theta$$

By using the transformation:

$$\text{let } z = \frac{n^0y^0}{\theta} \Rightarrow \theta = \frac{n^0y^0}{z}, |J| = \frac{n^0y^0}{z^2}$$

$$M_r = \frac{(n^0y^0)^{n^0+1}}{\Gamma(n^0+1)} \int_0^{\infty} \left(\frac{n^0y^0}{z}\right)^{-n^0-1-1+r} e^{-z}\frac{n^0y^0}{z^2} dz$$

$$M_r = \frac{(n^0y^0)^{n^0+1-n^0-2+r+1}}{\Gamma(n^0+1)} \int_0^{\infty} z^{n^0+2-r-2} e^{-z} dz$$

$$M_r = \frac{(n^0y^0)^r}{\Gamma(n^0+1)} \Gamma(n^0 - r + 1) \tag{14}$$

After that, we extract the posterior distribution through the Bayes rule, as shown in the following steps:

$$f(\theta\backslash t)=\frac{\frac{(n^0y^0)^{n^0+1}}{\Gamma(n^0+1)}\theta^{-(n^0+1)-1}e^{-\frac{n^0y^0}{\theta}}\left(\frac{\beta}{\theta}\right)^n\prod_{i=1}^{n}t_i^{\beta-1}e^{-\frac{\tau(t)}{\theta}}}{\int_0^\infty\frac{(n^0y^0)^{n^0+1}}{\Gamma(n^0+1)}\theta^{-(n^0+1)-1}e^{-\frac{n^0y^0}{\theta}}\left(\frac{\beta}{\theta}\right)^n\prod_{i=1}^{n}t_i^{\beta-1}e^{-\frac{\tau(t)}{\theta}}\,d\theta}$$

$$=\frac{\frac{(n^0y^0)^{n^0+1}}{\Gamma(n^0+1)}\beta^n\prod_{i=1}^{n}t_i^{\beta-1}\theta^{-(n^0+n+1)-1}e^{-\frac{(n^0y^0+\tau(t))}{\theta}}}{\frac{(n^0y^0)^{n^0+1}}{\Gamma(n^0+1)}\beta^n\prod_{i=1}^{n}t_i^{\beta-1}\int_0^\infty\theta^{-(n^0+n+1)-1}e^{-\frac{(n^0y^0+\tau(t))}{\theta}}d\theta}$$

$$=\frac{\theta^{-(n^0+n+1)-1}e^{-\frac{(n^0y^0+\tau(t))}{\theta}}}{\int_0^\infty\theta^{-(n^0+n+1)-1}e^{-\frac{(n^0y^0+\tau(t))}{\theta}}d\theta}=\frac{\theta^{-(n^0+n+1)-1}e^{-\frac{(n^0y^0+\tau(t))}{\theta}}}{\frac{\Gamma(n^0+n+1)}{(n^0y^0+\tau(t))^{n^0+n+1}}\int_0^\infty\frac{(n^0y^0+\tau(t))^{n^0+n+1}}{\Gamma(n^0+n+1)}\theta^{-(n^0+n+1)-1}e^{-\frac{(n^0y^0+\tau(t))}{\theta}}d\theta}$$

$$=\frac{(n^0y^0+\tau(t))^{n^0+n+1}}{\Gamma(n^0+n+1)}\theta^{-(n^0+n+1)-1}e^{-\frac{(n^0y^0+\tau(t))}{\theta}}\tag{15}$$

$$f(\theta\backslash t)\sim IG(n^0+n+1,n^0y^0+\tau(t))$$

After we get the posterior distribution and according to the above equation, we extract the standard deviation, as in the following steps:

$$M_r=\frac{(n^0y^0+\tau(t))^{n^0+n+1}}{\Gamma(n^0+n+1)}\int_0^\infty\theta^{-(n^0+n+1)-1+r}e^{-\frac{(n^0y^0+\tau(t))}{\theta}}\,d\theta$$

By using the transformation:

$$\text{let } z=\frac{(n^0y^0+\tau(t))}{\theta}\Rightarrow\theta=\frac{(n^0y^0+\tau(t))}{z},\ |J|=\frac{(n^0y^0+\tau(t))}{z^2}$$

$$M_r=\frac{(n^0y^0+\tau(t))^{n^0+n+1}}{\Gamma(n^0+n+1)}\int_0^\infty\left(\frac{n^0y^0+\tau(t)}{z}\right)^{-n^0-n-2+r}e^{-z}\frac{(n^0y^0+\tau(t))}{z^2}\,dz$$

$$M_r=\frac{(n^0y^0+\tau(t))^{n^0+n+1-n^0-n-2+r+1}}{\Gamma(n^0+n+1)}\int_0^\infty(z)^{n^0+n-r}e^{-z}\,dz$$

$$M_r=\frac{(n^0y^0+\tau(t))^r}{\Gamma(n^0+n+1)}\Gamma(n^0+n-r+1)\tag{16}$$

$$\text{s.d posterior}=\sqrt{\frac{(n^0y^0+\tau(t))^2}{(n^0+n)^2(n^0+n-1)}}\tag{17}$$

The above equation represents the standard deviation of the posterior distribution, after that the comparison is made between the value of the standard deviation of the prior distribution with the standard deviation of the posterior distribution. If the value of the standard deviation of the prior distribution is greater than the value of the standard deviation of the posterior distribution, this means that there is a problem of prior data conflict.

A second way to get the standard deviation of the posterior distribution is through the following steps:

Through the following form which represents the posterior distribution:

$$f(\theta/t) = \frac{\left(n^0y^0 + \tau(t)\right)^{n^0+n+1}}{\Gamma(n^0+n+1)}\theta^{-\left(n^0+n+1\right)-1}e^{-\frac{\left(n^0y^0+\tau(t)\right)}{\theta}}$$

Compensation for:

$$y^n = \frac{n^0y^0 + \tau(t)}{n^0+n}, n^n = n^0 + n$$

The posterior distribution becomes as follows:

$$f\left(\theta/n^ny^n\right) = \frac{\left(n^ny^n\right)^{n^n+1}}{\Gamma(n^n+1)}\theta^{-(n^n+1)-1}e^{-\frac{n^ny^n}{\theta}} \tag{18}$$

From the above, we conclude that $f(\theta/t) = f\left(\theta/n^ny^n\right)$ this means that the standard deviation of the prior distribution and the standard deviation of the posterior distribution will be according to the following formula:

$$\text{s.d prior} = \sqrt{\frac{\left(y^0\right)^2}{n^0-1}} \tag{19}$$

$$\text{s.d posterior} = \sqrt{\frac{\left(y^n\right)^2}{n^n-1}} \tag{20}$$

3.4 Standard error for mean

The indicator was employed for the purpose of testing the problem of prior data conflict. If the value of the standard error of the mean of the prior distribution is greater than the standard error of the mean of the posterior distribution, then this means that there is a problem of prior data conflict, and its formula is in the following form:

$$SE_{\bar{x}}.\text{prior} = \frac{\text{s.d prior}}{\sqrt{n}} \tag{21}$$

$$SE_{\bar{x}}.\text{posterior} = \frac{\text{s.d posterior}}{\sqrt{n}} \tag{22}$$

3.5 Address the problem of prior data conflict for Weibull distribution

Although this problem is represented by the prior data conflict problem, we can use a model to address the prior data conflict problem, This is done through the use of a set of prior parameters and according to the proposal presented by (Quaeghebeur and Cooman; 2005) [11], In short $\prod^0 = n^0 x \left[\underline{y}^0, \bar{y}^0\right]$, Another proposal was submitted (Walter and Augustin; 2009) In order to obtain a set of prior parameters, in brief $\prod^0 = [\underline{n}^0, \bar{n}^0] x \left[\underline{y}^0, \bar{y}^0\right]$, In general, the model presented to obtain a set of prior parameters is called (Generalized iLuck-Model), after that we get a set of posterior distributions, as shown below [3]:

$$f_1\left(\theta/\underline{n}^0, \underline{y}^0\right) = \frac{\left(\underline{n}^0\underline{y}^0\right)^{\underline{n}^0+1}}{\Gamma(\underline{n}^0+1)}\theta^{-(\underline{n}^0+1)-1}e^{-\frac{\underline{n}^0\underline{y}^0}{\theta}} \tag{23}$$

$$f_2\left(\theta/\underline{n}^0, \bar{y}^0\right) = \frac{\left(\underline{n}^0\bar{y}^0\right)^{\underline{n}^0+1}}{\Gamma(\underline{n}^0+1)}\theta^{-(\underline{n}^0+1)-1}e^{-\frac{\underline{n}^0\bar{y}^0}{\theta}} \tag{24}$$

$$f_3\left(\theta/\bar{n}^0, \underline{y}^0\right) = \frac{\left(\bar{n}^0\underline{y}^0\right)^{\bar{n}^0+1}}{\Gamma(\bar{n}^0+1)}\theta^{-(\bar{n}^0+1)-1}e^{-\frac{\bar{n}^0\underline{y}^0}{\theta}} \tag{25}$$

$$f_4\left(\theta/, \bar{n}^0, \bar{y}^0\right) = \frac{\left(\bar{n}^0\bar{y}^0\right)^{\bar{n}^0+1}}{\Gamma(\bar{n}^0+1)}\theta^{-(\bar{n}^0+1)-1}e^{-\frac{\bar{n}^0\bar{y}^0}{\theta}} \tag{26}$$

Since:
$\underline{n}^0$:Minimum.
$\bar{n}^0$: Maximum.
$\underline{y}^0$: Minimum.
$\bar{y}^0$: Maximum.

The above equations represent a set of the prior distributions obtained through the iLuck-Model, after that we extract the posterior set of distributions according to the following steps:

The first posterior distribution: From Eq. (23) and by using the Bayes rule, we get the first posterior distribution, as in the following equation:

$$f_1(\theta\backslash t) = \frac{\left(\underline{n}^0\underline{y}^0 + \tau(t)\right)^{\underline{n}^0+n+1}}{\Gamma(\underline{n}^0+n+1)}\theta^{-(\underline{n}^0+n+1)-1}e^{-\frac{\left(\underline{n}^0\underline{y}^0+\tau(t)\right)}{\theta}} \tag{27}$$

The above equation represents the first posterior distribution which is the Inverse Gamma distribution and by taking advantage of the properties of the Inverse Gamma distribution we get the central moments as in the following equation:

$$M_r = \frac{\left(\underline{n}^0\underline{y}^0 + \tau(t)\right)^r}{\Gamma(\underline{n}^0+n+1)}\Gamma\left(\underline{n}^0+n-r+1\right) \tag{28}$$

The second posterior distribution: From Eq. (24) and by using the Bayes rule we get the second posterior distribution as in the following equation:

$$f_2(\theta\backslash t) = \frac{\left(\underline{n}^0\bar{y}^0 + \tau(t)\right)^{\underline{n}^0+n+1}}{\Gamma(\underline{n}^0+n+1)}\theta^{-(\underline{n}^0+n+1)-1}e^{-\frac{\left(\underline{n}^0\bar{y}^0+\tau(t)\right)}{\theta}} \tag{29}$$

In short:

$$f_2(\theta \backslash t) \sim IG\left(\underline{n}^0 + n + 1, \underline{n}^0 \bar{y}^0 + \tau(t)\right)$$

The above equation represents the second posterior distribution, which is the Inverse Gamma distribution, and by taking advantage of the properties of the Inverse Gamma distribution we get the central moments as in the following equation:

$$M_r = \frac{\left(\underline{n}^0 \bar{y}^0 + \tau(t)\right)^r}{\Gamma(\underline{n}^0 + n + 1)} \Gamma\left(\underline{n}^0 + n - r + 1\right) \tag{30}$$

The third posterior distribution: From Eq. (25) and by using the Bayes rule, we get the third posterior distribution as in the following equation:

$$f_3(\theta \backslash t) = \frac{\left(\bar{n}^0 \underline{y}^0 + \tau(t)\right)^{\bar{n}^0 + n + 1}}{\Gamma(\bar{n}^0 + n + 1)} \theta^{-\left(\underline{n}^0 + n + 1\right) - 1} e^{-\frac{\left(n^0 \underline{y}^0 + \tau(t)\right)}{\theta}} \tag{31}$$

$$f_3(\theta \backslash t) \sim IG\left(\bar{n}^0 + n + 1, \bar{n}^0 \underline{y}^0 + \tau(t)\right)$$

The above equation represents the third posterior distribution which is the Inverse Gamma distribution and by taking advantage of the properties of the Inverse Gamma distribution we get the central moments as in the following equation:

$$M_r = \frac{\left(\bar{n}^0 \underline{y}^0 + \tau(t)\right)^r}{\Gamma(\bar{n}^0 + n + 1)} \Gamma\left(\bar{n}^0 + n - r + 1\right) \tag{32}$$

The fourth posterior distribution: From Eq. (26) and by using the Bayes rule we get the fourth posterior distribution as in the following equation:

$$f_4(\theta \backslash t) = \frac{\left(\bar{n}^0 \bar{y}^0 + \tau(t)\right)^{\bar{n}^0 + n + 1}}{\Gamma(\bar{n}^0 + n + 1)} \theta^{-\left(\underline{n}^0 + n + 1\right) - 1} e^{-\frac{\left(\bar{n}^0 \bar{y}^0 + \tau(t)\right)}{\theta}} \tag{33}$$

$$f_4(\theta \backslash t) \sim IG\left(\bar{n}^0 + n + 1, \bar{n}^0 \bar{y}^0 + \tau(t)\right)$$

The above equation represents the fourth posterior distribution which is the Inverse Gamma distribution and by taking advantage of the properties of the Inverse Gamma distribution we get the central moments as in the following equation:

$$M_r = \frac{\left(\bar{n}^0 \bar{y}^0 + \tau(t)\right)^r}{\Gamma(\bar{n}^0 + n + 1)} \Gamma\left(\bar{n}^0 + n - r + 1\right) \tag{34}$$

After taking the arithmetic mean of the posterior distributions we get the iLuck-Model [10]:

$$\underline{y}^n = \text{lower}\left(y^n\right) = \begin{cases} \dfrac{\bar{n}^0 \underline{y}^0 + \tau(t)}{\bar{n}^0 + n} & \text{if } \bar{\tau}(t) \geq \underline{y}^0 \\ \dfrac{\underline{n}^0 \underline{y}^0 + \tau(t)}{\underline{n}^0 + n} & \text{if } \bar{\tau}(t) < \underline{y}^0 \end{cases} \tag{35}$$

$$\bar{y}^n = \text{uper}(y^n) = \begin{cases} \dfrac{\bar{n}^0\bar{y}^0 + \tau(t)}{\bar{n}^0 + n} & \text{if } \bar{\tau}(t) \le \bar{y}^0 \\ \dfrac{\underline{n}^0\bar{y}^0 + \tau(t)}{\underline{n}^0 + n} & \text{if } \bar{\tau}(t) > \bar{y}^0 \end{cases} \tag{36}$$

Eqs. (35), (36) represent a generalized iLuck-Model, a model that represents the lower bound and the model that represents the upper bound is chosen based on the value of $\bar{\tau}(t)$ the estimator we obtain will be in the form of an interval. Therefore we will take the average for that period and from the above the posterior distribution will be in the following form:

$$f(\theta/n^m y^m) = \frac{(n^m y^m)^{n^m+1}}{r(n^m+1)} \theta^{-(n^m+1)-1} e^{-\frac{n^m y^m}{\theta}} \tag{37}$$

$$n^m = \frac{\text{lower}(n^n) + \text{uper}(n^n)}{2}, y^m = \frac{\text{lower}(y^n) + \text{uper}(y^n)}{2}$$

3.6 Robust Bayesian estimation for scale parameter for Weibull distribution

From Eq. (37) and by using the squared loss function, we get a Bayes estimator for the scale parameter as follows [12]:

$$E(\theta/n^m, y^m) = \hat{\theta}_{Rob}$$

$$\hat{\theta}_{Rob} = y^m \tag{38}$$

3.7 Robust Bayesian estimation for survival function for Weibull distribution

From Eq. (37) and by using the squared loss function we get a Bayes estimator for the survival function as follows [12]:

$$\hat{S}(t) = \int_0^\infty S(t)\, f(\theta/t)\, d\theta$$

$$\hat{S}(t) = \frac{(n^m y^m)^{n^m+1}}{r(n^m+1)} \int_0^\infty e^{-\frac{t^\beta}{\theta}} \theta^{-(n^m+1)-1} e^{-\frac{n^m y^m}{\theta}} d\theta$$

$$\hat{S}(t) = \frac{(n^m y^m)^{n^m+1}}{r(n^m+1)} \int_0^\infty \theta^{-(n^m+1)-1} e^{-\frac{(n^m y^m + t^\beta)}{\theta}} d\theta$$

By using the transformation:

$$\text{let } z = \frac{(n^m y^m + t^\beta)}{\theta} \Longrightarrow \theta = \frac{(n^m y^m + t^\beta)}{z}; |J| = | - \frac{(n^m y^m + t^\beta)}{z^2} |$$

$$\hat{S}(t) = \frac{(n^m y^m)^{n^m+1}}{r(n^m+1)} \int_0^\infty \left(\frac{(n^m y^m + t^\beta)}{z}\right)^{-(n^m+1)-1} e^{-z} \frac{(n^m y^m + t^\beta)}{z^2} dz$$

$$\hat{S}(t) = \frac{\left(n^m y^m\right)^{n^m+1}}{r(n^m+1)} \left(n^m y^m + t^\beta\right)^{-(n^m+1)} \int_0^\infty (z)^{n^m} e^{-z}\, dz$$

$$\hat{S}_{Rob}(t) = \left(\frac{n^m y^m}{n^m y^m + t^\beta}\right)^{n^m+1} \tag{39}$$

4. The experimental side of the Weibull distribution

In this section, Weibull distribution data will be generated by using the R program to estimate the scale parameter and the survival function in the prior data conflict, as shown in the following **Tables 1–4**:

Through **Tables 1** and **2** the simulation results showed that the robust Bayesian estimator is the best through the comparison standard (IMSE) in the case of prior data conflict. From the above, the robust Bayesian estimator will be applied to the real data to estimate the scale parameter and the survival function.

5. Real data description for Weibull distribution

Real data of a size of (15) for heart attack patients were collected from Al-Manathira General Hospital of the Najaf Health Department for the year 2018, as the time of admission of the patient to the hospital until discharge was recorded and that all of them were in a state of death upon discharge. This data is complete data, ti = (2,1,1,1,1,2,1,3,7,1,2,10,7,1,1), as these times are in days.

Model		**β = 2**		**n^0**			**Best**
	y^0		**Lower**		**Upper**		
	Lower	**Upper**	**2**		**5**		
			θ	**n**	**$\hat{pdf}$**	**$\hat{pdf}_{rob}$**	
1	1.5	2	1.5	10	0.008738	0.004018	Robust Bayesian
				20	0.004564	0.002699	
				40	0.002184	0.001634	
2	2	3	2	10	0.006128	0.002333	
				20	0.003421	0.001845	
				40	0.001539	0.001094	
3	2.5	4	2.5	10	0.004801	0.001671	
				20	0.002811	0.001428	
				40	0.001295	0.000895	

Table 1.
Integrated mean square error (IMSE) of the probability density function (pdf) for the Weibull distribution in the case of prior data conflict.

Model	$\beta = 2$			n^0			
					Lower	upper	
					2	5	
	y^0		θ	n	$\hat{S}(t)$	$\hat{S}_{rob}(t)$	Best
	Lower	Upper					
1	1.5	2	1.5	*10*	0.005332	0.002705	Robust Bayesian
				20	0.002894	0.001948	
				40	0.001570	0.001263	
2	2	3	2	*10*	0.004891	0.002070	
				20	0.002916	0.001729	
				40	0.001462	0.001120	
3	2.5	4	2.5	*10*	0.005310	0.002075	
				20	0.002802	0.001591	
				40	0.001412	0.001050	

Table 2.
Integrated mean square error (IMSE) of the survival function for the Weibull distribution in the case of prior data conflict.

B = 2		n^0		
y^0		Lower	Upper	
		6	8	
lower	upper	$\hat{\theta}_{rob}$	Standard deviation prior	Standard deviation posterior
16	20	16.10352	280.4339	242.4159

Table 3.
Estimation of the scale parameter of the Weibull distribution in the case of prior data conflict.

B = 2		n^0	
y^0		Lower	Upper
		6	8
lower	upper	t	$\hat{S}_{rob}(t)$
16	20	1	0.937227322
		2	0.772418759
		3	0.561586411
		7	0.050817125
		10	0.003284835

Table 4.
Estimation of the survival function of the Weibull distribution in the case of prior data conflict.

6. Binomial distribution

It is one of the discrete distributions in which the experiment can be repeated for (n) times so that the probability of success is (p) and the probability of failure is (1-p), so that the probability density function has the following formula [13]:

$$f(x) = \binom{n}{x} p^x (1-p)^{n-x}, x = 0, 1, \dots, n \tag{40}$$

and the average is:

$$E(x) = np \tag{41}$$

The variance is:

$$\sigma^2 = np(1-p) \tag{42}$$

6.1 Bayesian estimation for parameter p for binomial distribution

Suppose we have a sample that follows the Binomial distribution shown in the Eq. (40) and the appropriate prior distribution is the Beta distribution according to the following formula [13]:

$$f(p/a, b) = \frac{1}{\beta(\alpha, \beta)} p^{\alpha-1}(1-p)^{\beta-1}; 0 < p < 1 \tag{43}$$

To obtain the posterior distribution according to the Bayes rule as follows:

$$f(p/x) = \frac{f(x_1, x_2, \dots, x_n/p) f(p/\alpha, \beta)}{\int_0^1 f(x_1, x_2, \dots, x_n/p) f(p/\alpha, \beta) dp}$$

$$= \frac{\binom{n}{\sum_{i=1}^{n} x} p^{\sum_{i=1}^{n} x} (1-p)^{n-\sum_{i=1}^{n} x} \frac{1}{\beta(\alpha, \beta)} p^{\alpha-1}(1-p)^{\beta-1}}{\int_0^1 \binom{n}{\sum_{i=1}^{n} x} p^{\sum_{i=1}^{n} x} (1-p)^{n-\sum_{i=1}^{n} x} \frac{1}{\beta(\alpha, \beta)} p^{\alpha-1}(1-p)^{\beta-1} dp}$$

$$= \frac{p^{\sum_{i=1}^{n} x + \alpha - 1} (1-p)^{n-\sum_{i=1}^{n} x + \beta - 1}}{\int_0^1 p^{\sum_{i=1}^{n} x + \alpha - 1} (1-p)^{n-\sum_{i=1}^{n} x + \beta - 1} dp}$$

Since:

$$\sum_{i=1}^{n} x = s$$

$$= \frac{1}{\beta(\alpha + s, n + \beta - s)} p^{\alpha+s-1} (1-p)^{n+\beta-s-1} \tag{44}$$

From Eq. (44) and by using the squared loss function we get a Bayes estimator for parameter (p) as follows:

$$E(p/s) = \hat{p}$$

$$\hat{p} = \frac{\alpha + s}{n + \alpha + \beta} \tag{45}$$

6.2 Bayesian estimation for survival function for binomial distribution

Through Eq. (44) and by using the squared loss function we get a Bayes estimator for the survival function as follows [11] (**Figures 1** and **2**):

$$\hat{S}(t) = \int S(t) f(p/s) dp$$

$$f(p/s) = \frac{1}{\beta(\alpha + s, n + \beta - s)} p^{\alpha+s-1}(1-p)^{n+\beta-s-1}$$

$$S(t) = \sum_{j=t}^{n} \binom{n}{j} p^{j}(1-p)^{n-j}$$

$$\hat{S}(t) = \frac{1}{\beta(\alpha + s, n + \beta - s)} \sum_{j=t}^{n} \frac{n!}{j!(n-j)!} \int_{0}^{1} p^{\alpha+s+j-1}(1-p)^{n+\beta-s-1+n-j} dp$$

$$\hat{S}(t) = \frac{1}{\beta(\alpha + s, n + \beta - s)} \sum_{j=t}^{n} \frac{n!}{j!(n-j)!} \int_{0}^{1} p^{\alpha+s+j-1}(1-p)^{2n+\beta-s-j-1} dp$$

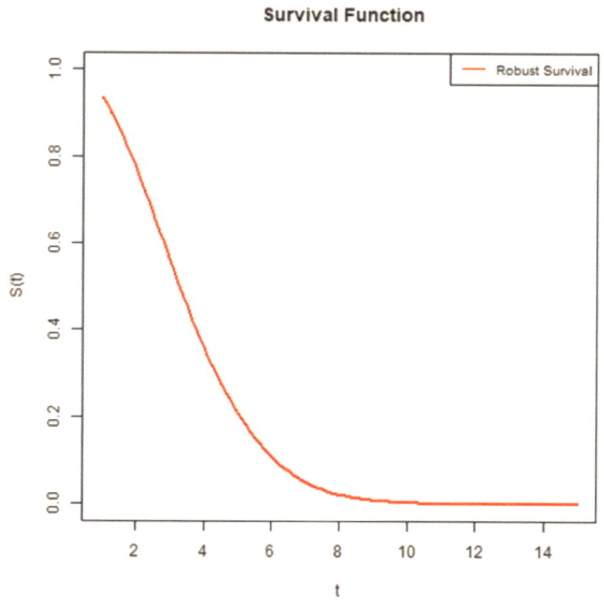

Figure 1.
It shows the behavior of the survival function by using the robust Bayes estimator, which is decreasing as the value of (t) increases, and this is consistent with the statistical theory [11].

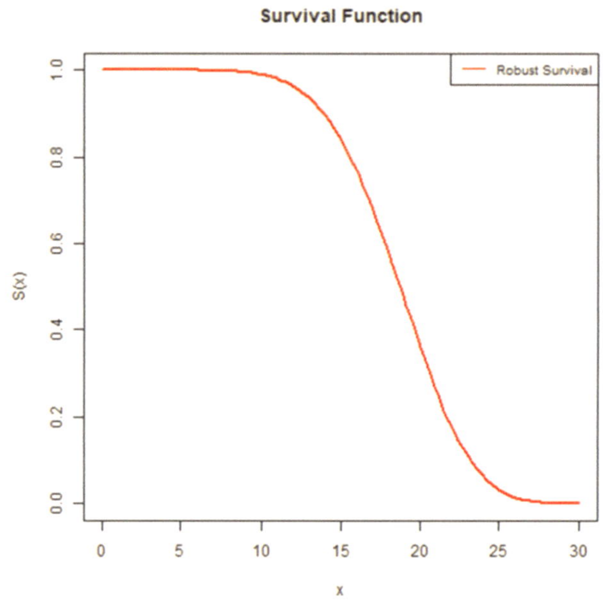

Figure 2.
It shows the behavior of the survival function of a binomial distribution which is decreasing and this is consistent with the statistical theory [13].

$$\hat{S}(t) = \frac{1}{\beta(\alpha + s, n + \beta - s)} \sum_{j=t}^{n} \frac{n!}{j!(n-j)!} \int_0^1 p^{\alpha+s+j-1}(1-p)^{2n+\beta-s-j-1} dp$$

Multiply and divide by:

$$\beta(\alpha + s + j, 2n + \beta - s - j)$$

$$\hat{S}(t) = \frac{1}{\beta(\alpha + s, n + \beta - s)} \sum_{j=t}^{n} \frac{n!}{j!(n-j)!} \frac{\beta(\alpha + s + j, 2n + \beta - s - j)}{\beta(\alpha + s + j, 2n + \beta - s - j)} \int_0^1 p^{\alpha+s+j-1}(1 - p)^{2n+\beta-s-j-1} dp$$

$$\hat{S}(t) = \frac{1}{\beta(\alpha + s, n + \beta - s)} \sum_{j=t}^{n} \frac{n!}{j!(n-j)!} \beta(\alpha + s + j, 2n + \beta - s - j) \tag{46}$$

6.3 Checking of prior data conflict for binomial distribution

Suppose that we have a sample that follows the binomial distribution and in short $x \sim$ bin (n,p) and as shown in the Eq. (40) then we need to determine the prior distribution through two methods [14]:

The first method: It is the Expected Conditional method we have previously explained this method so we go through the following steps [9]:

$$E(p/\alpha, \beta) = y^0 = \frac{\alpha}{\alpha + \beta}$$

$$y^0 = \frac{\alpha}{n^0}, \text{ where } n^0 = \alpha + \beta$$

$$y^0 = \frac{\alpha}{n^0} \Rightarrow \alpha = n^0 y^0, n^0 = n^0 y^0 + \beta \Rightarrow \beta = n^0 \left(1 - y^0\right)$$

Then we substitute the parameters $\alpha = n^0 y^0$, $\beta = n^0 \left(1 - y^0\right)$ with the prior distribution to get the prior distribution with the updated parameters:

$$f\left(p/n^0, y^0\right) = \frac{1}{\beta\left(n^0 y^0, n^0\left(1 - y^0\right)\right)} p^{n^0 y^0 - 1} (1 - p)^{n^0\left(1 - y^0\right) - 1} \tag{47}$$

The second method: In this method the prior distribution can be determined with the updated parameters through two steps, which are as follows [10]:

The first step: If the model can be written in the form of canonical exponential family, as shown below:

$$\begin{aligned} &f(x/p) = a(x) \exp\left(\psi.\tau(x) - nb(\psi)\right) \\ &f(s/p) = \binom{n}{s} p^s (1 - p)^{n-s} \\ &= \binom{n}{s} \exp\left\{ \ln\left(\frac{p}{1-p}\right) s - n(-\ln(1-p)) \right\} \\ &a(x) = \binom{n}{s}, \psi = \ln\left(\frac{p}{1-p}\right), \tau(x) = s, b(\psi) = -\ln(1-p) \end{aligned} \tag{48}$$

The second step: the prior distribution can be built by using the following model:

$$f\left(\psi/n^0, y^0\right) d\psi \propto \exp\left\{n^0 \left[y^0.\psi - b(\psi)\right]\right\} d\psi$$

$$f\left(\psi/n^0, y^0\right) d\psi \propto \exp\left\{n^0 \left[y^0 \ln\left(\frac{p}{1-p}\right) + \ln(1-p)\right]\right\} d\psi$$

$$\left|\frac{d\psi}{dp}\right| = \frac{1}{p(1-p)}$$

$$\begin{aligned} f\left(p/n^0, y^0\right) dp &= f\left(\psi/n^0, y^0\right) \left|\frac{d\psi}{dp}\right| dp \\ &\propto \exp\left\{n^0 y^0 \ln(p) + \left(n^0 - n^0 y^0\right) \ln(1-p)\right\} \frac{1}{p(1-p)} dp \end{aligned}$$

Then the prior distribution with the updated parameters is:

$$f\left(p/n^0, y^0\right) = \frac{1}{\beta\left(n^0 y^0, n^0\left(1 - y^0\right)\right)} p^{n^0 y^0 - 1} (1 - p)^{n^0\left(1 - y^0\right) - 1} \tag{49}$$

From the above equation, we extract the standard deviation of the prior distribution, as follows:

$$M_r = \frac{1}{\beta(n^0y^0, n^0(1-y^0))}\int_0^1 p^{n^0y^0+r-1}(1-p)^{n^0(1-y^0)-1}dp$$
$$= \frac{1}{\beta(n^0y^0, n^0(1-y^0))}\beta(n^0y^0+r, n^0(1-y^0))$$
$$= \frac{\Gamma(n^0y^0+n^0(1-y^0))}{\Gamma(n^0y^0)\Gamma(n^0(1-y^0))} * \frac{\Gamma(n^0y^0+r)\,\Gamma(n^0(1-y^0))}{\Gamma(n^0y^0+r+n^0(1-y^0))}$$
$$= \frac{\Gamma(n^0)\,\Gamma(n^0y^0+r)}{\Gamma(n^0y^0)\Gamma(n^0+r)} \tag{50}$$

$$\text{s.d prior} = \sqrt{\frac{y^0(1-y^0)}{n^0+1}} \tag{51}$$

From Eq. (47) and by using the Bayes rule, we extract the posterior distribution as shown in the following steps:

$$f(p/s) = \frac{\binom{n}{s}p^s(1-p)^{n-s}\frac{1}{\beta(n^0y^0, n^0(1-y^0))}p^{n^0y^0-1}(1-p)^{n^0(1-y^0)-1}}{\int_0^1\binom{n}{s}p^s(1-p)^{n-s}\frac{1}{\beta(n^0y^0, n^0(1-y^0))}p^{n^0y^0-1}(1-p)^{n^0(1-y^0)-1}\,dp}$$
$$= \frac{p^{n^0y^0+s-1}(1-p)^{n^0(1-y^0)+n-s-1}}{\int_0^1 p^{n^0y^0+s-1}(1-p)^{n^0(1-y^0)+n-s-1}dp}$$
$$= \frac{1}{\beta(n^0y^0+s, n^0(1-y^0)+n-s)}p^{n^0y^0+s-1}(1-p)^{n^0(1-y^0)+n-s-1} \tag{52}$$

Eq. (52) represents the posterior distribution, after that we extract the standard deviation of the posterior distribution as follows:

$$M_r = \frac{1}{\beta(n^0y^0+s, n^0(1-y^0)+n-s)}\int_0^1 p^{n^0y^0+s+r-1}(1-p)^{n^0(1-y^0)+n-s-1}\,dp$$
$$= \frac{1}{\beta(n^0y^0+s, n^0(1-y^0)+n-s)}\,\beta(n^0y^0+s+r, n^0(1-y^0)+n-s)$$
$$= \frac{\Gamma(n^0y^0+s+n^0(1-y^0)+n-s)}{\Gamma(n^0y^0+s)\Gamma(n^0(1-y^0)+n-s)} * \frac{\Gamma(n^0y^0+s+r)\Gamma(n^0(1-y^0)+n-s)}{\Gamma(n^0y^0+s+r+n^0(1-y^0)+n-s)}$$
$$= \frac{\Gamma(n^0+n)\,\Gamma(n^0y^0+s+r)}{\Gamma(n^0+n+r)\Gamma(n^0y^0+s)} \tag{53}$$

Since

$$f(p/s) = f(p/n^n, y^n)$$

Then the standard deviation of the posterior distribution is according to the following formula:

$$\text{s.d posterior} = \sqrt{\frac{y^n(1-y^n)}{n^n+1}} \tag{54}$$

6.4 Address the problem of prior data conflict for binomial distribution

In the part on the distribution of Weibull, how to solve this problem was explained, so we will enter the following steps [3]:

$$f_1(p/s) = \frac{1}{\beta\left(\underline{n}^0\underline{y}^0+s, \underline{n}^0\left(1-\underline{y}^0\right)+n-s\right)} p^{\underline{n}^0\underline{y}^0+s-1}(1-p)^{\underline{n}^0\left(1-\underline{y}^0\right)+n-s-1} \tag{55}$$

$$f_2(p/s) = \frac{1}{\beta\left(\underline{n}^0\bar{y}^0+s, \underline{n}^0(1-\bar{y}^0)+n-s\right)} p^{\underline{n}^0\bar{y}^0+s-1}(1-p)^{\underline{n}^0(1-\bar{y}^0)+n-s-1} \tag{56}$$

$$f_3(p/s) = \frac{1}{\beta\left(\bar{n}^0\underline{y}^0+s, \bar{n}^0\left(1-\underline{y}^0\right)+n-s\right)} p^{\bar{n}^0\underline{y}^0+s-1}(1-p)^{\bar{n}^0\left(1-\underline{y}^0\right)+n-s-1} \tag{57}$$

$$f_4(p/s) = \frac{1}{\beta\left(\bar{n}^0\bar{y}^0+s, \bar{n}^0(1-\bar{y}^0)+n-s\right)} p^{\bar{n}^0\bar{y}^0+s-1}(1-p)^{\bar{n}^0(1-\bar{y}^0)+n-s-1} \tag{58}$$

The first posterior distribution: From Eq. (55) and by using the Bayes rule, we get the posterior distribution, as in the following equation:

$$f_1(p/s) = \frac{1}{\beta\left(\underline{n}^0\underline{y}^0+s, \underline{n}^0\left(1-\underline{y}^0\right)+n-s\right)} p^{\underline{n}^0\underline{y}^0+s-1}(1-p)^{\underline{n}^0\left(1-\underline{y}^0\right)+n-s-1} \tag{59}$$

The above equation represents the first posterior distribution which is the Beta distribution and by using the properties of the Beta distribution we get the central moments, as in the following equation:

$$M_r = \frac{r(\underline{n}^0+n)\, r\left(\underline{n}^0\underline{y}^0+s+r\right)}{r(\underline{n}^0+n+r) r\left(\underline{n}^0\underline{y}^0+s\right)} \tag{60}$$

The second posterior distribution: From Eq. (56) and by using the Bayes rule, we get the second posterior distribution, as in the following equation:

$$f_2(p/s) = \frac{1}{\beta\left(\underline{n}^0\bar{y}^0+s, \underline{n}^0(1-\bar{y}^0)+n-s\right)} p^{\underline{n}^0\bar{y}^0+s-1}(1-p)^{\underline{n}^0(1-\bar{y}^0)+n-s-1} \tag{61}$$

The above equation represents the second posterior distribution, which is the Beta distribution, and by using the properties of the Beta distribution, we get the central moments, as in the following equation:

$$M_r = \frac{\Gamma(\underline{n}^0 + n)\,\Gamma\left(\underline{n}^0\bar{y}^0 + s + r\right)}{\Gamma(\underline{n}^0 + n + r)\Gamma\left(\underline{n}^0\bar{y}^0 + s\right)} \tag{62}$$

The third posterior distribution: From Eq. (57) and by using the Bayes rule, we get the third posterior distribution, as in the following equation:

$$f_3(p/s) = \frac{1}{\beta\left(\bar{n}^0\underline{y}^0 + s, \bar{n}^0\left(1 - \underline{y}^0\right) + n - s\right)} p^{\bar{n}^0\underline{y}^0 + s - 1}(1 - p)^{\bar{n}^0\left(1 - \underline{y}^0\right) + n - s - 1} \tag{63}$$

The above equation represents the third posterior distribution, which is the Beta distribution, and by using the properties of the Beta distribution, we get the central moments, as in the following equation:

$$M_r = \frac{\Gamma(\bar{n}^0 + n)\,\Gamma\left(\bar{n}^0\underline{y}^0 + s + r\right)}{\Gamma(\bar{n}^0 + n + r)\Gamma\left(\bar{n}^0\underline{y}^0 + s\right)} \tag{64}$$

The fourth posterior distribution: From Eq. (58) and by using the Bayes rule, we get the fourth posterior distribution, as in the following equation:

$$f_4(p/s) = \frac{1}{\beta\left(\bar{n}^0\bar{y}^0 + s, \bar{n}^0\left(1 - \bar{y}^0\right) + n - s\right)} p^{\bar{n}^0\bar{y}^0 + s - 1}(1 - p)^{\bar{n}^0\left(1 - \bar{y}^0\right) + n - s - 1} \tag{65}$$

The above equation represents the fourth posterior distribution, which is the Beta distribution, and by using the properties of the Beta distribution, we get the central moments, as in the following equation:

$$M_r = \frac{\Gamma(\bar{n}^0 + n)\,\Gamma\left(\bar{n}^0\bar{y}^0 + s + r\right)}{\Gamma(\bar{n}^0 + n + r)\Gamma\left(\bar{n}^0\bar{y}^0 + s\right)} \tag{66}$$

After taking the average of the posterior distributions, we get the iLuck-Model [10]:

$$\underline{y}^n = \text{lower}\left(y^n\right) = \begin{cases} \dfrac{\bar{n}^0\underline{y}^0 + \tau(x)}{\bar{n}^0 + n} & \text{if } \bar{\tau}(x) \geq \underline{y}^0 \\[2ex] \dfrac{\underline{n}^0\underline{y}^0 + \tau(x)}{\underline{n}^0 + n} & \text{if } \bar{\tau}(x) < \underline{y}^0 \end{cases} \tag{67}$$

$$\bar{y}^n = \text{uper}\left(y^n\right) = \begin{cases} \dfrac{\bar{n}^0\bar{y}^0 + \tau(x)}{\bar{n}^0 + n} & \text{if } \bar{\tau}(x) \leq \bar{y}^0 \\[2ex] \dfrac{\underline{n}^0\bar{y}^0 + \tau(x)}{\underline{n}^0 + n} & \text{if } \bar{\tau}(x) > \bar{y}^0 \end{cases} \tag{68}$$

Eqs. (67), (68) represent a generalized iLuck-Model, a model that represents the lower bound and the model that represents the upper bound is chosen based on the value of $\bar{\tau}(x)$, the estimator we obtain will be in the form of an interval Therefore, we will take the average for that period, and from the above, the posterior distribution model will be in the final form and as in the following equation:

$$f(p/n^m, y^m) = \frac{1}{\beta(n^m y^m, n^m(1-y^m))} p^{n^m y^m - 1}(1-p)^{n^m(1-y^m)-1} \tag{69}$$

$$n^m = \frac{lower(n^n) + uper(n^n)}{2}, y^m = \frac{lower(y^n) + uper(y^n)}{2}$$

6.5 Robust Bayesian estimation for parameter for binomial distribution

From Eq. (69) and by using the squared loss function, we get the robust Bayesian estimator for parameter (P) as follows [15]:

$$E(p/n^m, y^m) = \hat{p}_{Rob}$$

$$\hat{p}_{Rob} = \frac{n^m y^m}{n^m y^m + n^m(1-y^m)} \tag{70}$$

6.6 Robust Bayesian estimation for survival function for binomial distribution

From Eq. (69) and by using the squared loss function, we get a Bayesian estimator for the survival function, as follows [15]:

$$\hat{S}(x) = \int_0^1 S(x) f(p/s) dp$$

$$\hat{S}_{rob}(x) = \frac{1}{\beta(n^m y^m, n^m(1-y^m))} \sum_{j=x}^{n} \frac{n!}{j!(n-j)!} \int_0^1 p^j (1-p)^{n-j} p^{n^m y^m - 1}(1-p)^{n^m(1-y^m)-1} dp$$

$$\hat{S}_{rob}(x) = \frac{1}{\beta(n^m y^m, n^m(1-y^m))} \sum_{j=x}^{n} \frac{n!}{j!(n-j)!} \int_0^1 p^{n^m y^m + j - 1}(1-p)^{n^m(1-y^m)+n-j-1} dp$$

Multiply and divide the equation by:

$$\beta(n^m y^m + j, n^m(1-y^m) + n - j)$$

$$\hat{S}_{rob}(x) = \frac{1}{\beta(n^m y^m, n^m(1-y^m))} \sum_{j=x}^{n} \frac{n!}{j!(n-j)!} \frac{\beta(n^m y^m + j, n^m(1-y^m) + n - j)}{\beta(n^m y^m + j, n^m(1-y^m) + n - j)} \int_0^1 p^{n^m y^m + j - 1}(1-p)^{n^m(1-y^m)+n-j-1} dp$$

$$\hat{S}_{Rob}(x) = \frac{1}{\beta(n^m y^m, n^m(1-y^m))} \sum_{j=x}^{n} \frac{n!}{j!(n-j)!} \beta(n^m y^m + j, n^m(1-y^m) + n - j) \tag{71}$$

7. The experimental side of the binomial distribution

In this section, Binomial distribution data will be generated by using the R program to estimate the (P) parameter and the survival function in the prior data conflict, as shown in the following tables:

Through **Tables 5** and **6** the simulation results showed that the robust Bayesian estimator is the best through the comparison standard (IMSE) in the case of prior data conflict. From the above, the robust Bayesian estimator will be applied to the real data to estimate the (P) parameter and the survival function for the Binomial distribution.

Model	n^0							
	lower					upper		
	2					4		
	y^0		P	K	n	$\widehat{pmf}$	$\widehat{pmf}_{rob}$	Best
	Lower	upper						
1	0.2	0.4	0.4	5	10	0.006340	0.005353	Robust Bayesian Estimator
				10	20	0.002334	0.002143	
				20	40	0.000830	0.000799	
2	0.3	0.5	0.5	5	10	0.006121	0.005225	
				10	20	0.002207	0.002041	
				20	40	0.000832	0.000802	
3	0.4	0.6	0.6	5	10	0.005728	0.004898	
				10	20	0.002369	0.002206	
				20	40	0.000770	0.000743	

Table 5.
The integrated mean square error (IMSE) of the probability mass function (pmf) for the binomial distribution in the case of prior data conflict.

Model	n^0							
	Lower					upper		
	2					4		
	y^0		p	k	n	$\hat{S}(x)$	$\hat{S}_{rob}(x)$	Best
	Lower	Upper						
1	0.2	0.4	0.4	5	10	0.034387	0.027765	Robust Bayesian Estimator
				10	20	0.024367	0.022128	
				20	40	0.018268	0.017509	
2	0.3	0.5	0.5	5	10	0.034540	0.028961	
				10	20	0.024535	0.022408	
				20	40	0.016549	0.015884	
3	0.4	0.6	0.6	5	10	0.032064	0.026952	
				10	20	0.022817	0.021020	
				20	40	0.016745	0.016075	

Table 6.
The integrated mean square error (IMSE) of the survival function for the binomial distribution in the case of prior data conflict.

8. Real data description for the binomial distribution

Mortality data for patients with breast cancer were collected from Yarmouk Teaching Hospital for the period from 2010 to 2017, and the data collected are as follows (**Tables 7–9**):

Year	2010	2011	2012	2013	2014	2015	2016	2017
X_i	3	4	2	3	2	4	1	0

Table 7.
The real data for the binomial distribution.

y^0		n^0		
		Lower		upper
		4		6
Lower	Upper	$\hat{P}_{rob}$	Standard deviation prior	Standard deviation posterior
0.3	*0.6*	0.603595	0.042	0.006646

Table 8.
Robust Bayesian estimator for parameter (P) of the binomial distribution.

y^0		n^0	
		Lower	upper
		4	6
Lower	Upper	x	$\hat{S}_{rob}(x)$
0.3	*0.6*	0	1
		1	0.999
		2	0.996
		3	0.994
		4	0.986

Table 9.
Robust Bayesian estimator for survival function of the binomial distribution.

Author details

Ahmed Saadoon Mannaa
Statistics Department, College of Administration and Economics, University of Baghdad, Iraq, Baghdad

*Address all correspondence to: ahmed.sadoun1001@coadec.uobaghdad.edu.iq

References

[1] Quaehebeur E, Decoman G. Imprecise Probability Models for Inference in Exponential Symposium on Imprecise Probabilities and their Applications. Pittsburgh, Pennsylvania; 2005

[2] David JN, Michael E. Using Prior Expansions for Prior-Data Conflict Checking. 2020. Available from: https://arxiv.org/pdf/1902.10393

[3] Walter G, Augustin T. Imprecision and prior-data conflict in generalized Bayesian inference. Journal of Statistical Theory and Practice. 2009;**3**:255-271

[4] Abdulrahman S. Comparing different estimators of three parameters for transmuted Weibull distribution. Global Journal of Pure and Applied Mathematics. 2017;**13**:5115-5128

[5] Abdulabaas F, Al-Mayali Y, Neama I. A comparison between the Bayesian and the classical estimators of Weibull distribution. Journal of Kufa for Mathematics and Computer. 2013;**1**(8): 21-28

[6] Erlandson F, Adriano K. Bayesian computational methods for estimation of two-parameters Weibull distribution in presence of right-censored data. Chilean Journal of Statistics. 2017;**8**(2):25-43

[7] Chris B, Ibrahim A, Al OM. Bayesian estimation of two-parameter Weibull distribution using extension of Jeffrey's prior information with three loss functions. Mathematical Problems in Engineering. 2012;**2**:13-28

[8] Guure C, Ibrahim N, Adam M. Bayesian parameter and reliability estimate of Weibull failure time distribution. Bulletin of the Malaysian mathematical sciences. Society. 2014; **37**(3):27

[9] Walter G, Coolen F. Robust Bayesian reliability for complex systems under prior-data conflict. ASCE-ASME Journal of Risk and Uncertainty in Engineering Systems. 2018;**4**. DOI: 10.1061/AJRUA6.0000974

[10] Walter G. Generalized Bayesian Inference with Sets of Conjugate Priors for Dealing with Prior-Data Conflict Course at Lund University. Lund, Sweden: Lund University; 2015

[11] Al-Nasser AM. An Introduction to Statistical Reliability. Amman, AL-Sharjha, Al-Khabor: UB Group, Ithraa Publishing and Distribution, University Book Shop, Elmia Book Stores; 2009

[12] Al-Doori EA, Mannaa AS. Robust Bayesian estimators for survival function under prior data conflict with practical application in the health side. Indian Journal of Forensic Medicine & Toxicology. 2020;**14**(2):778-789

[13] Razzaghi M. On the estimation of binomial success probability with zero occurrence in sample. The Journal of Modern Applied Statistical Methods. 2002;**17**(1):28-40

[14] Walter G, Coolen P. Sets of Priors Reflecting Prior-Data Conflict and Agreement. Switzerland: Springer International Publishing; 2016. DOI: 10.1007/978-3-319-40596-4_14

[15] Al-Doori EA, Mannaa AS. Robust Bayesian estimators for binomial distribution under prior data conflict. Periodicals of Engineering and Natural Sciences. 2020;**8**(1):284-297

Section 3

Bayesian Applications

Chapter 4

Applications of Hierarchical Bayesian Methods to Answer Multilayer Questions with Limited Data

Frederick Bloetscher

Abstract

There are many types of problems that include variables that are not well defined. Seeking answers to complex problems that involve many variables becomes mathematically challenging. Instead, many investigators use methods like principal component analysis to reduce the number of variables, or linear or logistic regression to rank the impact of the variables and eliminating those with the limited impact. However, eliminating variables can create a loss of integrity, especially for variables that might be associated with low likelihood but have high impact events. The use of hierarchical Bayesian methods resolves this issue by utilizing the benefits of information theory to help answer questions by incorporating a series of prior distributions for a number of variables used to solve an equation. The concept is to create distributions for the range and likelihood for each variable, and then create additional distributions to define the mean and shape values. At least three levels of analysis are required, but the hierarchical solution can include added levels beyond the initial variables (i.e., distributions related to the priors for the shape parameters). The results incorporate uncertainty, variability, and the ability to update the confidence in the values of the variables based on the receipt of new data.

Keywords: predictive Bayesian, hierarchical, Drake, infrastructure, dose–response, risk, extreme events

1. Introduction

Suppose you have a complex question with numerous variables that are not well understood. The challenge that confronts us all is that such situations are not unusual—there are many examples of situations where there are numerous variables that can contribute to occurrences in our world, whether these occurrences involve medical issues, infrastructure issues, or science questions. Statistical methods have been developed to address limited information, but most do not permit the incorporation of new information except Bayesian methods.

The development of Bayesian methods that include priors that can be updated with new data or can respond as a result of added data overcomes initial limitations of most models. Bayesian methods developed as a result of information theory, assuming that the absolute or unconditional probability density function $p(x)$ on X is the underlying distribution found through curve-fitting. Priors can be determined based on any combination of subjective or numeric information in the absence of real data, or as data are collected, including parameters such as the mean, variance, and range. Utilization of the observations from the prior data leads to the posterior probability function, which incorporates observations from x in the sample space S, although revealing additional information about the true content of the sample space S is subject to the influence of the proper prior distribution assumptions for x [1–3].

Many Bayesian practitioners stop with the posterior function, but the ability to develop true statistical inference requires further effort to create a predictive Bayesian solution. The Bayesian posterior methods were used in prior studies [2–12]. Predictive Bayesian methods are an extension of traditional Bayesian approaches, in which unconditional distributions for the quantity of interest are found by integrating over probabilities of parameters of the distribution for the quantity of interest, incorporating both uncertainty and variability in the quantity of interest. They have been termed "believed probabilities."

Press [13] noted that there are advantages to the predictive Bayesian approach. Practical experience and subjectivity can be accounted for explicitly by fitting known or subjective data to a probability function that can be updated as added information becomes available [13]. Predictive Bayesian methods continually improve the statistical inference based on increased amounts of data (hence more data should advance the understanding of statistical relationships and provide greater confidence in the prior and therefore the solution). The predicted distributions are also important for checking goodness of fit of the resulting predictive model to actual data. However, the analysis can become problematic when the information is so scarce that the analysis yields nothing useful [11].

The use of Monte Carlo methods makes the solutions easier. Through randomized sampling, the resulting predictions are simulated from the posterior predictive distribution, which is the distribution of the unobserved future results based on prior observed data. The more confidence that exists with the priors, the more likely results of likely outcomes can be derived.

However, many times Bayesian methods have been limited to situations where there are one or two variables that contribute to an outcome. While high-quality answers can be derived, the design of the algorithm often oversimplifies the real world where many variables may contribute to the outcome. The ability to incorporate many variables that are unknown or uncertain makes the calculations intractable in the traditional Bayesian processes.

Equally important is the ability to study the events, where multiple variables may impact the probability and impact of a given consequence. Regression models are often used, along with principal component analysis to address such situations. However, both rely on complete data sets, and for many situations, the lack of complete data may be extensive. Examples include much of the public or municipal infrastructure that we rely on so heavily for a functioning society, health risks or impacts, natural disaster risk, and extreme event prediction. This is where hierarchical application to Bayesian methods has value.

2. Methods

Multiple variables create greater challenges. The concept is to create a distribution with shape and location parameters for each variable, then create distributions to define the mean and shape values (priors). Each of those distributions can be solved with MCMC methods to create a predictive distribution. That predictive distribution can be sampled and multiplied (or added) to the results of all other variables solved and sampled similarly. The assignment of distributions need not stop with the priors for shape location and location. Those parameters also can be assigned a distribution, and likewise those (and so on). Model parameters create a structure or a series of levels that look like a hierarchy, whereby the priors of a given solution are dependent on priors to those variable distributions and integrate that information across levels simultaneously [12], thereby separating the observed variability into parts attributable to both random and true differences [13–15]. Each investigation will have a different series of variables, each with different associated variables, priors, and priors of the priors.

Allenby et al. [16] stated that hierarchical Bayesian models are really "the combination of two things: i) a model written in hierarchical form that is ii) estimated using Bayesian methods." Shaddick et al. [17] consider there are to be at least three levels: (1) the observation or measurement level, (2) the underlying process level, and (3) the parameter level. Kruschke and Vanpaemel [18] noted that hierarchical Bayesian data analysis involves "describing data by meaningful mathematical models and allocating credibility to parameter values that are consistent with the data and with prior knowledge." Using Bayesian methods, hierarchical Bayesian models can yield estimates of the true effects at each level of the hierarchy [14, 19]. By considering the results across all levels, hierarchical Bayesian models can be used to rigorously integrate information with a complex underlying structure [14], resulting in a tendency to shrink differences when multiple variables are incorporated [20]. An important aspect of the hierarchical approach is that the model is usually a flexible version of a base model [21], and if needed, the models allow for adding extra levels depending on the hyperparameters [22].

Applying Bayesian prediction and weighting in a unified approach to Bayesian regression models can account for complex design features under the framework of multilevel regression and poststratification [23–25]. Weighting in a hierarchical model can be used as an extension of linear or logistic regression models. Methods for hierarchical functional data typically require that all curves are observed over or standardized to fall in the same region [26–28]. While classical weighting usually relies on many user-defined choices for regression that is difficult to codify [29], the Bayesian approach allows prior information to be incorporated and the distributions automatically adjusted [30, 31].

MCMC allows the user to approximate aspects of posterior distributions that cannot be directly calculated (e.g., random samples from the posterior, posterior means, etc.). There are examples of current applications of this approach. Draper [32] considered Bayesian hierarchical Poisson regression models, Wang et al. [33] created hierarchical Bayesian model developed for predicting monthly residential per capita electricity consumption at the state level across the United States, and Maddala et al. [34] studied the relationship of income elasticity on energy demand in the United States by applying a dynamic linear regression model under Bayesian framework. Roman et al. [35] and Neil and Fenton [36] used hierarchical Bayesian model for evaluation of treatments for Covid-19.

There are limits to hierarchical Bayesian model. The first is the underlying resulting model assumption may be wrong [14, 31]. Thais et al. [37] noted that "ill behaved likelihoods" at the lower levels in the hierarchy may create either excessive concentration about a mean or noninformative results. Rouder et al. [38] note that although hierarchical linear models are suitable in several domains, they rarely make good models of psychological process. Heller and Gharamani [39] note the algorithm provides no guide to choosing the "correct" number of clusters.

As a result, Shaddick et al. [18] note that Bayesian hierarchical models are an extremely useful and flexible framework in which to model complex relationships and dependencies in data, while Kruschke and Vanpaemel [17] suggest they provide flexibility in designing models that are appropriate for describing the data at hand that can provide a complete representation of parameter uncertainty (i.e., the posterior distribution) that can be directly interpreted. Some examples on how to create a hierarchical Bayesian model are helpful to demonstrate the process.

3. Applications

3.1 Example 1: Drake equation

Since 1959, SETI has yet to find an alien signal. Two questions arise as a result—what is the probability of there being life in the galaxy, and why have not we received a response to our transmissions? In 1959, a US astronomer, Frank D. Drake, a NASA employee who carried out the first SETI radio telescope experiments, outlined an equation for finding communicable civilizations [40, 41]:

$$N = RL \tag{1}$$

Which was later expanded to : $Rf_p n_e f_i f_t f_c L$.

where N is the number of communicable civilizations, R is the rate at which stars are born in the galaxy, f_p is the fraction of stars with planetary systems, n_e is the number of planets that might hold life, f_t is the fraction of planets with life, f_i is the fraction of planets with life that have evolved, f_c is the number of civilizations of evolved civilizations with the ability to communicate, and L is the length of time over which the communication is possible. An additional factor named C is a recent suggestion for colonization [42]. Bloetscher [3] suggested that the factor n_e actually comprises four factors: planet size (PS), presence of a moon (M), location within the "Goldilocks" or habitable zone (HZ), and the correct star type (ST), creating four unknowns from one. None of the 7–12 factors is fully known, so no specific answer on the likelihood of intelligent life on another planet communicating with Earth is possible. However, hierarchical Bayesian methods can be used to investigate the probability of intelligent life on another planet communicating with Earth. This approach involves the assignment of probability distributions to the underlying factors and using those to develop an MCMC protocol to determine the final predictive solution [3]. Subjective data, shown in **Table 1**, were included when little or no data were available to specify the parameters of these distributions [3]. Then probability distributions were assigned to the prior parameters within the initial distributions to determine the location and scale parameters of the factor distribution (**Table 2**). The subjective information serves to create these prior distributions until such time as real data are developed or become available [3].

Variable description	Source of variable	Variable	Drake and Sobel (1991) estimate	Diehl et al. (2006) or Maccone (2010) estimate	Bloetscher mean from literature	Range of potential values	Distribution type	Prior1 distribution	Prior 1 estimate	Prior2 distribution	Prior 2 estimate	Data points
Number of Starts forming each year	Drake and Soble, 1991	R	10	7	20.0	0 to infinity	Gamma	Gamma	4	Gamma	0.2	Numerous, but actually only 1
Percent of stars with planets	Drake and Soble, 1991	fp	0.5	0.5	0.7	0 to 1	Beta	Uniform	1	Uniform	1.4	1
Number of Planets with life	Drake and Soble, 1991	ne	2	1	n/a	0 to infinity	Gamma	Gamma	1	Gamma	0.125	1
Percent of stars of right type	Bloetscher, 2019	ST	n/a	n/a	0.7	0 to 1	Beta	Uniform	1.4	Uniform	2	Thousands, and we find them with planets
Percent of planets of the right size	Bloetscher, 2019	PS	n/a	n/a	0.25	0 to 1	Beta	Uniform	1	Uniform	4	8
Percent of planets in Habitable zone	Bloetscher, 2019	HZ	n/a	n/a	0.375	0 to 1	Beta	Uniform	1	Uniform	2.33	8
Percent of planets with moons to create motion	Bloetscher, 2019	M	n/a	n/a	0.5	0 to 1	Beta	Uniform	1	Uniform	2	Dozens
Percent of planets where life forms	Drake and Soble, 1991	fl	1	0.5	0.5	0 to 1	Beta	Uniform	0.5	Uniform	0.5	1
Percent of planets with life that becomes intelligence	Drake and Soble, 1991	fi	0.01	0.2	0.5	0 to 1	Uniform	n/a		n/a		1

Variable description	Source of variable	Variable	Drake and Sobel (1991) estimate	Diehl et al. (2006) or Maccone (2010) estimate	Bloetscher mean from literature	Range of potential values	Distribution type	Prior1 distribution	Prior 1 estimate	Prior2 distribution	Prior 2 estimate	Data points
Percent of planets that develop technology to communicate	Drake and Soble, 1991	fc	0.01	0.2	0.2	0 to 1	Beta	Uniform	0.5	Uniform	0.5	1
Lifetime of civilization that wants to communicate	Drake and Soble, 1991	L	10,000	10,000	4600	0 to infinity	Exponential	Gamma	1	n/a		1
Percent of civilizations with a desire to colonize	Walters et al. 1980	C	n/a	n/a	0.2	0 to 1	Beta	Uniform	1	Uniform	5	1
	Mean of N		10	700	21							

Table 1.
Summary of Drake parameters used in Bayesian calculations and comparison to prior estimates.

Parameter	Mean (10,000-samples)
R	31.89
fp	0.4392
ST	0.4433
PS	0.2586
HZ	0.3170
M	0.3768
fl	0.5013
fi	0.5016
fc	0.4956
L	4641
C	0.2219
MCMC | Average for N	24.64

Table 2.
Monte Carlo results for parameters used in MCMC.

For the Drake equations, a distribution for *N* was developed through using the Hierarchical Monte Carlo distributions for the factors of the equation run 10,000 times (see **Figure 1**). A series of Hierarchical Monte Carlo algorithms were developed for each parameter, and the means were inserted into a Monte Carlo Markov Chain program that uses a Metropolis-Hastings algorithm with a Gibbs sampler to develop a final probabilistic result [43–49] that was solved for *N*. Based on suggestions by Glade et al. [50] and Maccone [51], the target MCMC distribution was proposed to be log-normal. Given the uncertainly involved, the standard deviation used for the target distribution was assumed to be the square root of 6, after Wu et al. [52]. Given a multivariate distribution, like the example above, Gibbs sampling breaks down the problem by drawing samples for each parameter directly from that parameter's

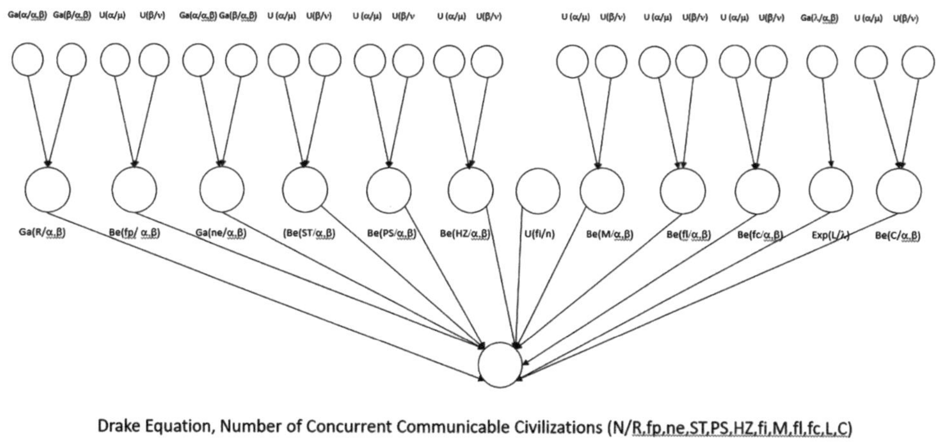

Figure 1.
Development of HPB for Drake.

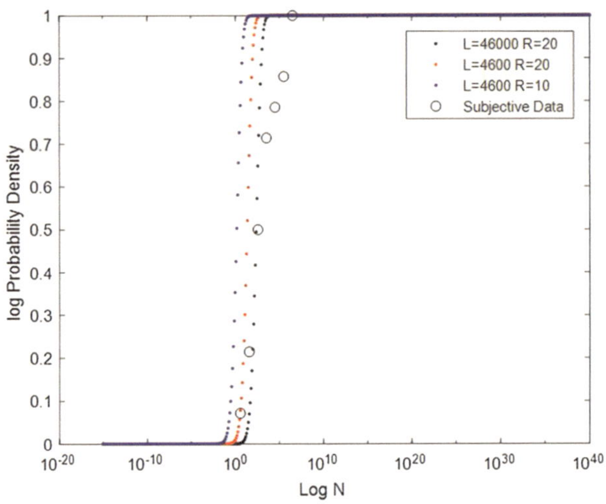

Figure 2.
Results for the probability of N *using the Monte Carlo calculation of the factors placed into a MCMC predictive Bayesian MATLAB calculation. Note that the number of other planets is constantly small despite three options as noted above (reproduced from Bloetcher [3]).*

conditional distribution or the probability distribution of a parameter *given* a specific value of another parameter [53].

The solution is shown in **Figure 2** (red data points). Of importance, there is nearly a 50% probability that we are alone in the galaxy. The graph indicates that there is a 95% probability that there are less than 100 communicating civilizations concurrent with Earth, and a 99% that there are 1000 such civilizations.

3.2 Example 2: dose response

The use of predictive Bayesian methods for dose–response relationships has also been investigated by a number of authors [10–12, 54–56]. Beaudequin et al. [57] developed QMRA with the use of hierarchical Bayesian networks to address the data paucity, combine quantitative and qualitative information including expert opinion, and the ability to offer a systems approach to characterize complexity. They outlined how the Bayesian networks are the current method of choice for determining the risk to human health from exposure to pathogens because of their ability to separate risk and uncertainly, predict outcomes, and deal with poorer quality data [57]. Hence, as subjective data are incorporated, the prior distributions self-adjust [58]. Bloetscher et al. [9] used six sets of *Cryptosporidium* data to show how the dose–response function changes with new, additional data. As a result of new data, the dose–response is expected to improve, demonstrating that the process can be applied to other organisms. In addition, the paper creates a Predictive Bayesian MCMC solution for the Pareto II distribution with two uncertain parameters.

Given the unlikelihood of reinfection during a single incident (due to a short period of time), the likelihood of infection can be described by the binomial distribution. As such, a binomial function is used to represent the probability of exposure. **Figure 3** shows the conceptual model with three levels of probability distributions.

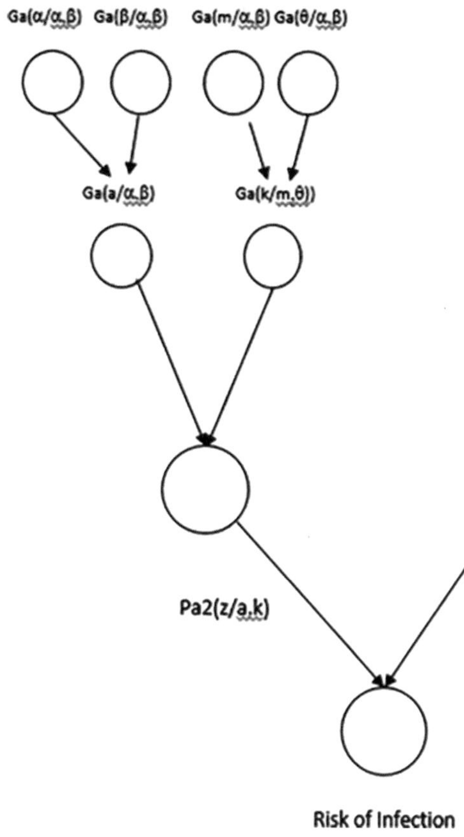

Figure 3.
Conceptual risk model for cryptosporidium.

Because of the intrinsic difficulty in solving a predictive Bayesian equation with multiple embedded distributions through double integration, an analytical mathematical solution is not achievable. Instead, a probabilistic solution was developed using a Markov Chain Monte Carlo (MCMC) program developed in MATLAB18® with uncertain values for a and k. Six different models of 10,000 iterations were run, each model including an additional dataset and the prior for α increased to account for the additional data.

When compared with the beta-Poisson models developed by Haas et al. [59], the predictive Bayesian equation derived in this study is less conservative by a factor of over 10 than the beta-Poisson model used by Haas et al. [59] (see **Figure 4**). However, the beta-Poisson does not accept new data, and therefore cannot be updated, is the likely explanation for the difference.

3.3 Example 3: infrastructure

Public water and sewer utility systems are created to develop safe, reliable, and financially self-supporting potable water and sanitary sewage systems, which will meet the water and sewerage needs of the areas served by the utility, to ensure that existing and future utility facilities are constructed, operated, and managed with high reliability and are compatible with the area's future growth. To gain efficiencies in

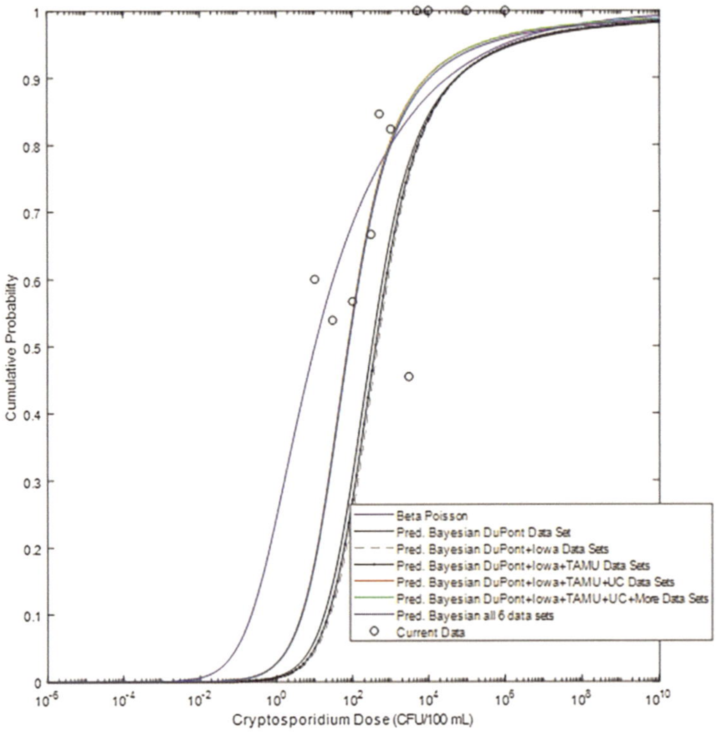

Figure 4.
Dose–response function for C. Parvum*—All six datasets compared with beta-Poisson. The beta-Poisson is the most conservative of the sets. Once the last three datasets (TAMU, UC, and More) are added; all provide the same graph (reproduced from Bloetscher, et al. 2020).*

operation, these new facilities must be developed in accordance with the latest technical and professional standards to protect the health, safety, and welfare of the citizens served now or in the future.

Public infrastructure has been poorly rated by the American Society of Civil Engineers [60–65], and most public officials acknowledge the deterioration of the infrastructure we rely on daily. Part of the challenge is that many jurisdictions have limited information about their systems, and little data to use to justify spending of specific project. Hence the infrastructure tends to deteriorate further each year as local officials opt to limit budgets in the absence of good needs data. As a result, state and local governments currently spend about 1.8% of its GNP on infrastructure, as compared with 3.1% in 1970 [66]. Twice as much was spent 40 years ago, and a large portion of today's costs are for growth as opposed to repair and replacement. Asset management is supposed to help this meet this challenge.

An asset management program consists of determining the selected area of study, type of system, and the quality of data used for evaluation. The question is how to collect data that might be useful to a utility that does not involve a lot of destructive testing on buried infrastructure that is costly and inconvenient. When creating an asset management plan, missing data are perceived to be a huge problem, especially when the event data (breaks in pipe as an example) are not tracked. The lack of tracking makes it difficult to determine which factors are the critical ones. Many utilities lack the resources for examining buried infrastructure, so other methods of data collection are needed. The concept in Bloetscher et al. [67] was to develop a means to acquire data on the assets for a condition assessment (buried pipe is not visible and cannot really be assessed). What was found was that for buried infrastructure, much more information was known than anticipated. For one thing, most utilities have a pretty good idea about the pipe materials. Employee memory can be very useful, even if not completely accurate. In most cases, the depth of pipe is fairly similar—the deviations may be known. Soil conditions may be useful—there is an indication that aggressive soil causes more corrosion in ductile iron pipe, and most soil information is readily available. Groundwater is usually known, and if a saltwater interface of a pollution plume exists, it can be mapped and evaluated for impact on pipe. Tree roots will wrap around water and sewer pipes, so their presence is detrimental. Trees are easily noted from aerial photographs. Roads with heavy truck traffic create more vibrations in the soil, causing rocks to move toward the pipe and joints to flex. So, with a little research, there are at least six variables known.

All variable information can be compiled into tables. There is also a need to track events or consequences—breaks, flooding etc.—that would indicate a failure, which is required for predicting future maintenance needs and the most at-risk assets. Finally, the data along with the consequence can be used to predict where the breaks might occur in the future based on past experience. If the break history for a water system, flood records for a stormwater system, or sewer pipe condition from televising is known, the impact of these factors can be developed via a linear regression algorithm. For logistic of linear regression, XLStat® can be used for the statistical analysis. The linear regression algorithm can then be used as a predictive tool to help identify assets that are mostly likely to become a problem.

Data need to be kept up as things change, but exact data are not needed. An example of this type of effort is shown for a medium-sized city in Florida in **Figures 5–7**. The City's GIS system was mined for the purposes of this project. Data were retrieved and reviewed to address missing data and clear errors. Nearly 10,000 pipe segments remained. Categorical information on trees, vibrations, soil type, and pipe type is added. Noncategorical data for pipe size, length, and age were also entered. Note that with 10,000 pipe sections and less than 600 breaks, many

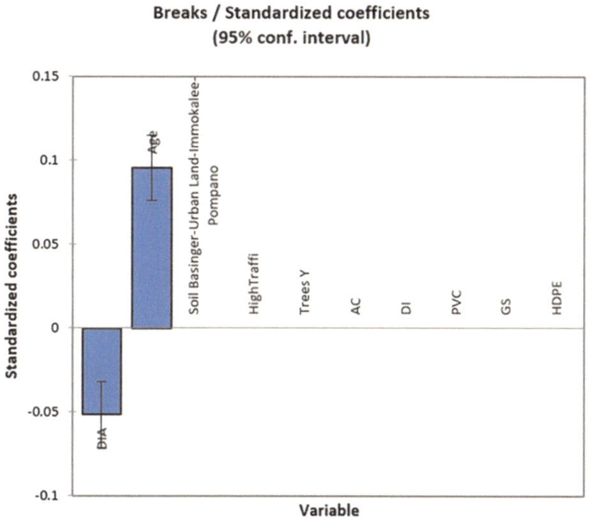

Figure 5.
Impact of factors on leaks.

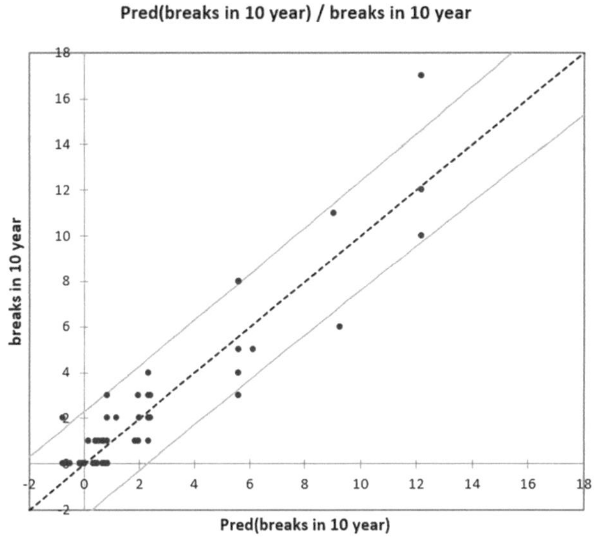

Figure 6.
Comparison of predictive and actual breaks over 10 years (correlation desirable).

pipes have no breaks in their history. The linear regression function for XLStat® was used to create equation to identify the factors associated with each variable and the amount of influence that each exerts (see **Figure 5**). In this case, the equation was:

$$\text{Breaks} = -3.54427\text{E} - 03 - 6.5187\text{E} - 03 * \text{DIA} + 2.607\text{E} - 03 * \text{Age} \qquad (2)$$

It should be noted that this utility has three main types of pipe, installed at three completely different eras. Because the correlation between pipe type and age was high, and likewise pipe type and diameters, other factors that might impact leaks in other communities were not obvious, so other communities would need to recreate this analysis for their situation.

Figure 6 outlines how the predictive equation correlated for the City's potable water distribution system (well within one standard deviation). **Figure 7** is a GIS map of pipe vulnerability based on the data. Red pipe is the highest priority to schedule for replacement.

The concept should apply to any utility, although the results and factors of concern will be slightly different for each utility. Also, in smaller communities, many variables (ductile iron pipe, PVC pipe, soil condition ...) may be so similar that attempts to differentiate factors may be unproductive.

The analysis indicated two things—that age and AC pipe were correlated.

But what if none of this information is fully known? Many of the indicators of failure can be tracked through the information that is required to be included in the as-built drawings, but what if they are not available? Loss on institutional knowledge through retirements can cost the utility much information on actual pipe diameter, pipe depth, age, and breaks given many utilities do not have extensive work order systems. Other information that might be useful is condition that maintenance crews may have knowledge of. A hierarchical Bayesian model could be developed to address these concerns. Where the pipe is actually known, the categorical variable would be

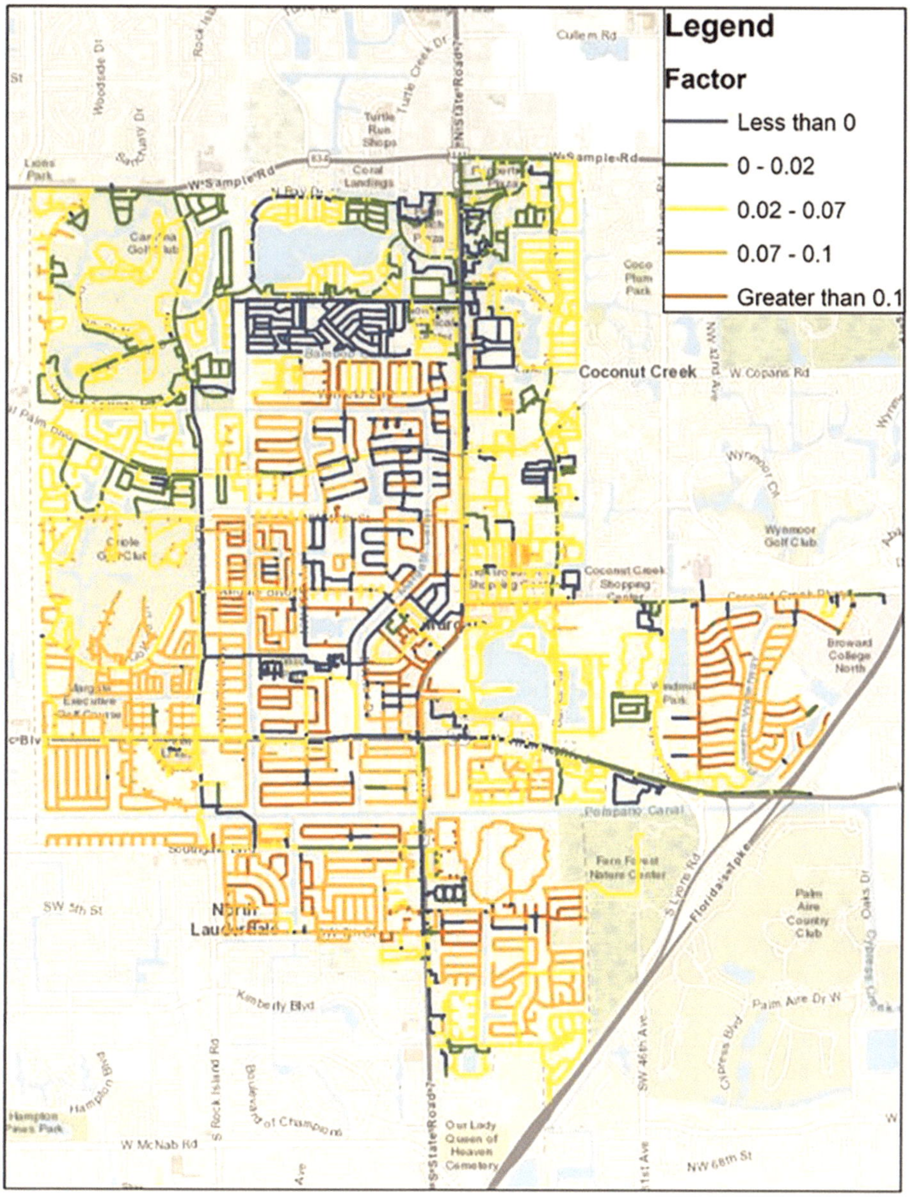

Figure 7.
Pipe risk—Red pipe is the highest risk for this community, while blue pipes are the lowest risk.

set to 1. Otherwise, a beta distribution could be developed with a "confidence mean"—we think it is ductile iron, but it might be PVC or cast iron. The same with pipe diameters, etc. As new variables are developed, confidence could be added and priors adjusted. Criticality could be a distribution as well. **Figure 8** shows what an infrastructure assessment Hierarchical Predictive Bayesian model might look like (realizing it might extend far more widely). Currently, research is underway to develop such models, but the data required to create and utilize the models are often lacking even in the most sophisticated organizations.

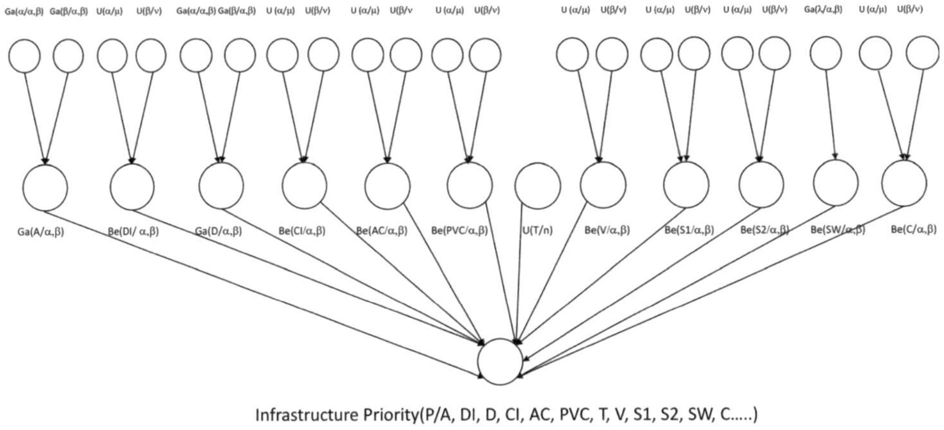

Figure 8.
Partial diagram of potential hierarchical predictive Bayesian infrastructure risk model.

4. Conclusions

Going back to the beginning of the chapter, "What to do when you have a complex question with numerous variables that are not well understood?" It would appear that the use of hierarchical predictive Bayesian models is a solution to address the challenge. While there may be circumstances where these methods may not work (psychology), for issues such as infrastructure and completely unknown questions such as the Drake equation, the methods seem ideally situated to shed light on the solution in a probabilistic form. The outcome of these methods provides a probability of a given answer—not a specific answer—at different levels of confidence. Uncertainty and variability are by the nature of being a probabilistic answer already included in the solutions. This is why added data can improve the likelihood of a given solution while reducing the potential for less likely solutions. Like the Drake equation solution, the dose–response example is an example of such a process. Infrastructure would be as well—as more data are created, the solutions become more robust and uncertainty is reduced. The results permit us to make better decisions as the data improve our understanding.

Author details

Frederick Bloetscher
Florida Atlantic University, USA

*Address all correspondence to: fbloetsc@fau.edu

References

[1] Aitchison J, Dunsmore IR. Statistical Prediction Analysis. Cambridge, UK: Cambridge University Press; 1975

[2] Bloetscher F, Englehardt JD, Chin DA, Rose JB, Tchobanoglous G, Amy VP, et al. Comparative assessment municipal wastewater disposal methods in Southeast Florida. Water Environment Research. 2005;**77**: 480-490

[3] Bloetscher F. Using predictive Bayesia n Monte Carlo- Markov Chain methods to provide a probablistic solution for the Drake equation. Acta Astronautica. 2019;**155**:118-130. DOI 10.1016/j. actaastro.2018.11.033

[4] Englehardt JD, Lund J. Information theory in risk analysis. Journal of Environmental Engineering, American Society of Civil Engineers. 1992;**118**(6):890-904. DOI 10.1061/(ASCE)0733-9372(1992) 118:6(890)

[5] Englehardt JD. Scale invariance of incident size distributions in response to sizes of their causes. Risk Analysis, Society for Risk Analysis. 2002;**22**(2): 369-381

[6] Englehardt JD. Response: Pareto incident size distribution. ASCE Journal of Environmental Engineering. 1997; **123**(1):99-101

[7] Englehardt JD. Predicting incident size from limited information. Journal of Environmental Engineering. 1995; **121**(5):455-464

[8] Englehardt JD. Pollution prevention technologies: A review and classification. Journal of Hazardous Materials. 1993;**35**: 119-150. DOI 10.1016/0304-3894(93) 85027-C

[9] Bloetscher F, Meeroff DE, Long SC, Dudle JD. Demonstrating the benefits of predictive Bayesian dose-response relationships using 6 exposure studies of cryptosporidium Parvum. Risk Analysis. 2020;**40**(11):2442-2461. DOI 10.1111/ risa.13552

[10] Bloetscher F, Meeroff DE, Phonpornwithoon P. Assessing risk of injection of reclaimed water into the Biscayne aquifer for aquifer recharge purposes. Journal of Geoscience and Environment Protection. 2019;**07**: 184-201. DOI 10.4236/gep.2019.77013

[11] Bloetscher F. Development of a Predictive Bayesian Microbial Dose-Response Function, Doctoral Dissertation. Coral Gables, FL: University of Miami; 2001

[12] Englehardt JD, Swartout P. Predictive population dose-response assessment for cryptosporidium Parvum: Infection endpoint. Journal of Toxicology and Environmental Health Part A: Current Issues. 2004;**67**(8–10): 651-666

[13] Press SJ. Bayesian Statistics: Principles, Models and Applications, John Wiley & Sons, Inc., New York, NY Gelman A, Stern HS, Carlin JB, Dunson DB, Vehtari A, Rubinb DB. 2013. Bayesian Data Analysis. 3rd ed. Boca Raton, FL: CRC Press; 1989

[14] McGlothlin AE, Viele K. Bayesian hierarchical models. JAMA. 2018; **320**(22):2365

[15] Zhai C, Lafferty J. In: Croft W, Harper D, Kraft D, Zobel J, editors. Document Language Models, Query Models, and Risk Minimization for Information Retrieval. SIGIR Conference

on Research and Development in Information Retrieval. New York: ACM Press; 2001. pp. 111-119

[16] Allenby GM, Rossi PE, McColloch RE. Hierarchical Bayes Models: A practitioners Guide. 2004. https://www.semanticscholar.org/paper/Hierarchical-Bayes-Models-Allenby-Rossi/322edbc740ecd21e1e3b454e3c9a09deb3d11e39 [Accessed: 2/22/22]

[17] Shaddick G, Green M, Thomas M. Bayesian Hierarchical Models University of Bath, Symposium, 6th - 9th December 2016. 2016

[18] Kruschke JK, Vanpaemel W. Bayesian estimation in hierarchical models. In: Busemeyer JR, Wang Z, Townsend JT, Eidels A, editors. The Oxford Handbook of Computational and Mathematical Psychology. Oxford, UK: Oxford University Press; 2015. pp. 279-299

[19] Quintana M, Viele K, Lewis RJ. 2017. Bayesian analysis: Using prior information to interpret the results of clinical trials. JAMA. 2017;**318**(16): 1605-1606. DOI 10.1001/jama.2017.15574

[20] Lipsky AM, Gausche-Hill M, Vienna M, Lewis RJ. 2010. The importance of "shrinkage" in subgroup analyses. Annals of Emergency Medicine. 2010;**55**(6):544-552. DOI 10.1016/j.annemergmed.2010.01.002

[21] Simpson D, Rue H, Riebler A, Martins TG, Sorbye SH. Penalising model component complexity: A principled, practical approach to constructing priors. Statistical Science. 2017;**32**:1-28

[22] Chib S, Greenberg E. In: Durlauf SN, Lawrence E, editors. Hierarchical Bayes Models in the New Palgrave Dictionary of Economics. 2nd ed. Chicago, IL: Blume; 2008

[23] Gelman A, Little TC. Post-stratification into many categories using hierarchical logistic regression. Survey Methodology. 1997;**23**:127-135

[24] Ghitza Y, Gelman A. Deep interactions with MRP: Election turnout and voting patterns among small electoral subgroups. American Journal of Political Science. 2013;**57**:762-776

[25] Park DK, Gelman A, Bafumi J, J. State-level opinions from national surveys: Poststratification using multilevel logistic regression. In: Cohen JE, editor. Public Opinion in State Politics. Redwood City, CA: Stanford University Press; 2005

[26] Brumback B, Rice J. Smoothing spline models for the analysis of nested and crossed samples of curves. JASA. 1998;**93**:961-976

[27] Brumback L, Lindstrom M. Self modeling with flexible, random time transformations. Biometrics. 2004;**60**: 461-470

[28] Morris J, Vannucci M, Brown P, Carroll R. Wavelet-based nonparametric modeling of hierarchical functions in colon carcinogenesis. JASA. 2003;**98**:573-583

[29] Gelman A. Struggles with survey weighting and regression modeling. Statistical Science. 2007;**22**:153-164

[30] Si Y, Trangucci R, Gabry JS, Gelman A. Bayesian hierarchical weighting adjustment and survey inference. 2017. https://arxiv.org/abs/1707.08220. [Accessed: 2/22/22]

[31] Ghosh M, Meeden G. Bayesian Methods for Finite Population Sampling. Boca Raton, FL: CRC Press; 1997

[32] Draper D. Discussion of the paper by lee and Nelder. Journal of the Royal Statistical Society, Series B. 1996;**58**: 662-663

[33] Wang S, Sun X, Lall U. A hierarchical Bayesian regression model for predicting summer residential electricity demand across the U.S.a. Energy. 2017;**140**(2017):601-611

[34] Maddala GS, Trost RP, Li H, Joutz F. Estimation of short-run and long-run elasticities of energy demand from panel data using shrinkage estimators. Journal of Business & Economic Statistics. 1997; **15**:90-100

[35] Roman YM, Burela PA, Pasupuleti V, et al. Ivermectin for the treatment of COVID-19: A systematic review and meta-analysis of randomized controlled trials. Clinical Infectious Diseases. 2021:ciab591. DOI 10.1093/cid/ciab591

[36] Neil M, Fenton N. Bayesian hypothesis testing and hierarchical Modeling of Ivermectin effectiveness author information risk information and management research, School of Electronic Engineering and Computer Science, Queen Mary University of London, London, United Kingdom the authors have no conflicts of interest to declare. American Journal of Therapeutics. 2021;**28**(5):e576-e579. DOI 10.1097/MJT.0000000000001450

[37] Thaís C, Fonseca O, Migon HS, Mirandola H. Reference Bayesian Analysis for Hierarchical Models. Ithaca, NY: Cornell University; 2019. Available online: https://arxiv.org/abs/1904.11609v1

[38] Rouder JN, Morey RD, Pratte MS. September 2, 2013 Hierarchical Bayesian Models. 2013. http://pcl.missouri.edu/sites/default/files/p5.pdf. [Accessed: 2/22/22]

[39] Heller KA, Gharamani Z. ND. Bayesian Hierarchical Clustering, bhcnew.dvi (ucl.ac.uk). [Accessed: 2/2/22]

[40] Drake FD. Discussion of Space Science Board, National Academy of Sciences Conference on Extraterrestrial Life, Nov 1961. WV: Green bank; 1961

[41] Jones BW. SETI: The search for extraterrestrial intelligence. Physics Education. 1991;**26**:52-57

[42] Walters C, Hoover RA, Kotra RK. Interstellar colonization: A new parameter for the Drake equation. Icarus. 1980;**41**(2):193-197

[43] Besag J, Green PJ, Higdon D, Mengersen KLM. Bayesian computation and stochastic systems (with discussion). Statistical Science. 1995;**10**:3-66

[44] Casella G, George EI. Explaining the Gibbs sampler. The American Statistician. 1992;**46**:167-174

[45] Hastings WK. Monte Carlo sampling methods using Markov chains and their applications. Biometrika. 1970; **57**:97-109

[46] Haugh M. MCMC and Bayesian Modeling, IEOR E4703 Monte-Carlo Simulation. New York, NY: Columbia University; 2017

[47] Metropolis N, Ulam S. The Monte Carlo method. Journal of the American Statistical Association. 1949; **44**:335-341

[48] Metropolis N, Rosenbluth AW, Rosenbluth MN, Teller A, Teller H. Equations of state calculations by fast computing machines. Journal of Chemical Physics. 1953;**21**:1087-1091

[49] Walsh 2002. Markov Chain Monte Carlo and Gibbs Sampling Lecture Notes for EEB 596z. http://nitro.biosci.arizona.edu/courses/EEB596/handouts/Gibbs.pdf

[50] Glade N, Ballet P, Bastien O. A stochastic process approach of the drake equation parameters. International Journal of Astrobiology. 2011;**11**(2): 103-108. DOI 10.1017/S1473550411000413

[51] Maccone C. The statistical Drake equation. Acta Astronautica. 2010;**67**: 1366-1383

[52] Wu Z-N, Li J, Bai C-Y. Scaling relations on log Normal type growth process with an extremal principle of entropy. Entropy. 2017;**19**:56. DOI 10.3390/e19020056

[53] Smith AFM, Roberts GO. Bayesian computation via the Gibbs sampler and related Markov chain Monte Carlo methods. Journal of the Royal Statistical Society: Series B (Methodological). 1993; **55**(1):3-23. DOI 10.1111/j.2517-6161.1993.tb01466.x

[54] Teunis PFM, Havelaar AH. The beta Poisson dose-response model is not a single-hit model. Risk Analysis. 2000;**20**: 513-520

[55] Teunis PF, Ogden ID, Strachan NJ. Hierarchical dose response of E. coli O157:H7 from human outbreaks incorporating heterogeneity in exposure. [research support, non-U.S. Gov't]. Epidemiology and Infection. 2008; **136**(6):761-770. DOI 10.1017/S0950268807008771

[56] Englehardt JD, Swartout P. Predictive Bayesian microbial dose-response assessment based on suggested self-Organization in Primary Illness Response: Cryptosporidium parvum. Risk Analysis. 2006;**26**(2):651-666. DOI 10.1111/j.1539-6924.2006.00745.x

[57] Beaudequin D, Harden F, Roiko A, Stratton H, Lemckert C, Mengersen K. Beyond QMRA: Modelling microbial health risk as a complex system using Bayesian networks. Environment International. 2015;**80**:8-18. DOI 10.1016/j.envint.2015.03.013

[58] Johnson NL, Kotz S. Continuous Univariate Distributions I. New York, NY: Wiley and Sons; 1970

[59] Haas C, Rose J, Gerba C. Quantitative microbial risk assessment. New York: John Wiley & Sons; 1999

[60] ASCE 2001. 2001 Report Card for America's Infrastructure. http://ascelibrary.org/doi/book/10.1061/9780784478882. [Accessed: 2/22/22]

[61] ASCE 2005. 2005 Report Card for America's Infrastructure. http://ascelibrary.org/doi/book/10.1061/9780784478851. [Accessed: 2/22/22]

[62] ASCE 2009. 2009 Report Card for America's Infrastructure, ASCE, Alexandria. http://www.infrastructurereportcard.org/making-the-grade/report-card-history/2001-report-card/. [Accessed: 2/22/22]

[63] ASCE 2013. 2013 Report Card for America's Infrastructure, ASCE, Alexandria. http://www.infrastructurereportcard.org/. [Accessed: 2/22/22]

[64] ASCE 2017. 2017 Report Card for America's Infrastructure, ASCE, Alexandria. http://www.infrastructurereportcard.org/making-the-grade/report-card-history/2001-report-card/. [Accessed: 2/22/22]

[65] ASCE. Report card on America's Infrastructure, ASCE. 2021. https://infrastructurereportcard.org/. [Accessed: 2/22/22]

[66] McNichol D. The Roads That Built America: The Incredible Story of the US Interstate System. New York, NY: Sterling; 2006

[67] Bloetscher F, Wander L, Smith G, Dogon N. Public infrastructure asset assessment with limited data. Open Journal of Civil Engineering. 2017, 2017; **07**(03):79326. 20 pages. DOI 10.4236/ojce.2017.73032

Chapter 5

Bayesian Inference as a Tool to Optimize Spectral Acquisition in Scattering Experiments

Alessio De Francesco, Luisa Scaccia, Martin Bohem and Alessandro Cunsolo

Abstract

Nowadays, an increasing number of scattering measurements rely on the use of large-scale research facilities, which is usually granted after highly competitive peer-reviewing and typically for short-time lapses. The optimal use of the allocated time requires rigorous estimates on the reliability of the data analysis, as inferred from the limited statistical accuracy of the measurement. Bayesian inference approaches can significantly help this endeavor by providing investigators with much-needed guidance under challenging decisions on experimental time management. We propose here a method based on the real-time data analysis of running experiments, which fully exploits the core strengths of Bayes theorem. The procedure is implemented in sequential steps in which the spectral measurement is adjourned by summing to it successive acquisition runs, and the spectral modeling is upgraded accordingly. At each stage, the statistical accuracy of the measurement improves, and a more grounded joint posterior distribution is drawn and used as a prior in the subsequent data acquisition stage. The gradual reduction in the model parameters' uncertainty down to the targets set *a priori* by experimenters provides a quantitative "success criterion," which helps prevent oversampling during acquisition. A similar "on the fly" data modeling, might substantially change the way large-scale facilities operate.

Keywords: Bayesian inference, neutron and X-ray scattering, spectroscopy, MCMC methods, Bayes theorem, Brillouin neutron scattering

1. Introduction

Nowadays, fundamental and applied research in Condensed Matter Physics relies heavily on the use of large research infrastructures. These include continuous or spallation neutron sources or X-rays synchrotron facilities present today worldwide. Often, sources of neutrons and X-rays are indeed found in the same geographical place, given the recognized complementarity of these two powerful spectroscopic techniques for the study of matter. This is the case of the European Photon and

Neutron (EPN) campus in Grenoble, France, which hosts the Institut Laue-Langevin (ILL) [1], and the European Synchrotron Research Facility (ESRF) [2]. Similarly, both neutron and X-ray facilities are hosted by the Rutherford Appleton Laboratory in Oxfordshire, UK, (ISIS and Diamond, respectively) [3, 4], by the Paul Sherrer Institut in Villigen, Swiss (SINQ and SLS) [5, 6], and soon by the city of Lund, Sweden (ESS and Max IV Laboratory, respectively) [7, 8]. Large-scale facilities are accessible to scientists for beam-time allocation through a highly competitive proposal selection carried out by expert panels through peer-review processes. Based on this peer-review outcome, the number of days (or even hours) assigned to an experiment is thoroughly pondered. It readily appears how critical is to establish an optimal experimental strategy enabling to gather the most informative and precise data out of an approved measurement. For this purpose, one needs to evaluate not only the ideal number of samples and related physical and chemical conditions, but frequently (if not always, in neutron scattering experiments) the time needed for a certain number of ancillary measurements that are mandatory to achieve a clean set of data. These include accurate measurements of the resolution function, the background signal, and spurious intensity effects, in which the raw measurement needs to be precisely corrected for. Therefore, an optimal use of the beam time assigned to an experiment would greatly benefit from a quantitative criterion to take sensible decisions during the measurement. Here, we propose a simple method to achieve such a criterion based on Bayesian statistics and its inferential capabilities [9–11]. In Section 2, we briefly describe an inelastic neutron or X-ray measurement and the main concerns rising when deciding its duration. In Section 3, we focus on the output of a Brillouin Neutron scattering (BNS) experiment: the spectrum of density fluctuations of a system; in particular, we show how one can use a Bayesian approach to model this observable. In the same section, we recall a fundamental property of the Bayes theorem that makes it suited to a recursive use for data analysis purposes. To demonstrate the potentialities of this approach, we reproduce the results of a typical BNS measurement by generating simulated experimental spectra. We then summarize the results of an on-the-fly data modeling of these spectra, which enables us to draw a joint posterior distribution for the adopted model parameters eventually guiding the decision on when conveniently stop a spectral acquisition.

Such a running analysis should establish the premises for developing a Measurement Integration Time Optimizer (MITO), a computational tool to assist scattering experiments in large-scale research facilities. In Section 4, we will shortly mention aspects of the approach described which deserves attention or caution; finally, in Section 5, conclusions and possible perspectives are outlined.

2. Neutron and X-ray scattering measurements

The main outcome of neutron and X-ray scattering measurements is the rate $\dot{N}$ of neutrons or photons scattered at an angle 2Θ with energy changed by an amount $\hbar\omega$, and ultimately captured by an array of detectors intercepting a finite solid angle. Aside of instrumental factors such as flux, detector efficiency, or detector sensitive area, the intensities recorded by the detectors depend on the physicochemical properties of the sample *via* its spectrum of density fluctuation, $S(Q, E)$ [12], which conveys insights on the structure (positions) and the dynamics (movements) of the atoms in the sample.

Obviously, the longer the detector counting, the more accurate the spectral shape determination. In fact, the number of neutron (x-ray) counted within an integration time t, $\dot{N}t$, obeys to a Poisson distribution, its standard deviation thus being $\sqrt{N}$. As the integration time t increases, the counting statistics improves as the relative experimental errors ($\sim 1/\sqrt{N} \equiv 1/\sqrt{\dot{N}t}$) decreases. Hence, the chance to detect interesting details of the spectral shape ultimately depends, of course, on the sample properties, but also on the accuracy of the intensity measurement. A difficulty to be faced in typical INS spectral acquisitions is that the measurement might be terminated prematurely, that is, before providing the information sought for. This possibility appears especially penalizing if the counting statistics achieved is not accurate enough, the spectral features not well-defined, or the signal sought for very weak. Conversely, data can also be integrated longer than strictly needed to capture the effect under scrutiny. In this case, further prolonging the counting would not complement the insight of the measurement significantly, and, beyond some time lapse, would not even improve its quality. Even worse, it could jeopardize the ultimate success of the experiment due to the time waste, which could prevent the accomplishment of the full experimental plan.

Without digging into computational details, here we outline a strategy to support experiments with a Bayesian protocol providing useful assistance in the measurement's planning and decision making. This will help investigators to determine when the integration time of a spectral acquisition can be safely stopped, either because all useful information was gathered, or because the predetermined target established for relative uncertainties was reached. For the sake of simplicity, we will focus on an exemplary inelastic neutrons scattering (INS) case, with the implicit assumption that the method can be safely extended to X-rays scattering (IXS), in fact being generally valid for any spectroscopy measurement. We stress that the case we are considering is very likely also the most demanding in terms of computational effort for reasons that will be briefly illustrated later in this chapter.

3. Inelastic neutron scattering

As mentioned, the general aim of an INS measurement is to measure the spectrum of density fluctuations $S(Q,E)$, which conveys insights on positions and movements of the atoms in a sample. Oversimplifying, depending on the spectrometer we use to determine it, we can have access to different aspects of $S(Q,E)$, either relating to collective movements of atoms (e.g., acoustic waves, structural relaxation processes), single-particle ones (e.g., translational diffusion, rotations, librations ...), or both. To measure $S(Q,E)$ of a given system one can use two different types of neutron spectrometers: triple axis spectrometers (TAS) and time-of-flight (TOF) ones.

Here, we assume to execute measurements with a TOF spectrometer [13] where $S(Q,E)$ surfaces are sampled, ideally, for each Q and E values simultaneously [14]. The rate of neutrons scattered at the different scattering angles 2Θ (see Appendix A) and impinging on the sensitive area of the detector after a time (of flight) t defines the time-dependent intensity function $I(2\Theta,t)$. The latter is converted into an intensity $I(Q,E)$, which is a function of the momentum, Q, and the energy, $E = \hbar\omega$, exchanged between sample and probe particles, with $\hbar$ and ω being the reduced Planck constant and the exchanged frequency, respectively. In Appendix A, a sketch of the BRISP spectrometer and few hints about the principles of the TOF technique are shortly recalled.

Aside from instrumental effects such as energy resolution and signal background, $I(Q,E)$ is proportional to $S(Q,E)$, which is the physical variable, INS (and IXS) investigators usually seek for. To sample and gather this intensity function with adequate counting statistics, providing us the needed information, a certain acquisition time is required, which depends on the characteristics of the instruments (incident neutron flux, detector efficiency, resolution ...) and on the scattering properties of the sample. These are embodied in its double differential cross section $d^2\sigma/d\Omega dE$ [12, 14] defined as the number of neutrons deviated in a second into the small solid angle $\Delta\Omega$ subtended by a detector along the 2Θ direction, with final energy included in the interval between E and $E + \Delta E$ [14]. More explicitly:

$$\dot{N} = \frac{d^2\sigma}{d\Omega dE} J \Delta\Omega\Delta E. \tag{1}$$

where J represents the incident flux of neutrons.

To summarize, the *image* of the $S(Q, E)$ is being built up as the measurement runs, and the larger the acquisition time, the more precise is the $S(Q, E)$ rendering. To avoid data loss of the entire measurement in case of instrument failure, the measurement is usually split into different sub-runs which, for the sake of simplicity we will hereafter assume to have the same acquisition time.

3.1 A Brillouin neutron scattering measurement

Once the $S(Q, E)$ surface is measured, different constant Q cuts of it are determined by interpolation. As an example of a typical INS measurement outcome, in **Figure 1** we show the spectrum of liquid silver at $Q = 6nm^{-1}$ measured with the Brillouin spectrometer BRISP [15] at the High Flux Reactor of the Institut Laue Langevin (Grenoble, France) [16, 17]. The spectral intensity has the typical shape of the spectrum from a disordered samples, which consists of a central peak broadened around the elastic energy $E = 0$ sided by a pair of inelastic peaks shifted from the elastic energy by an amount of $\pm\hbar\omega_s$, which defines the energy of the excitation. The

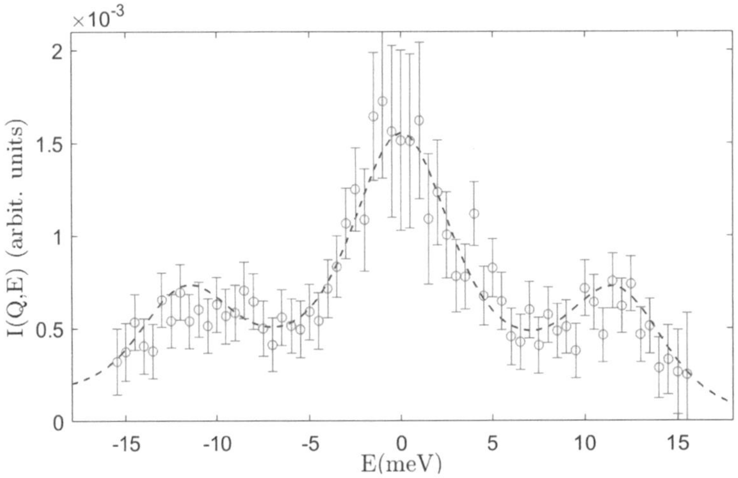

Figure 1.
Dynamic structure factor of liquid silver measured on the Brillouin spectrometer BRISP at a momentum transfer $Q = 6\ nm^{-1}$ and an incident energy $E_0 = 83.9\ meV$. The experimental data points are affected by resolution broadening. The dashed dot line is a fit obtained from an oversimplified model: A central Lorentzian line is summed to a damping harmonic oscillators (DHO) function to describe the inelastic excitations [16].

peaks sitting at positive and negative energy are respectively the well-known Stokes and Anti-Stokes lines of a Brillouin spectrum [18].

3.2 Modeling through a Bayesian approach

Let us assume that measured data can be described by a chosen model specified by the vector $\Theta = (\theta_1, \theta_2, \cdots, \theta_m)$, whose generic component θ_m is a model parameter. For the sake of generality, we include the possibility that some Θ components, instead of being the parameter of a well-identified model, designate instead one model option among several competitive ones whose reliability is to be concurrently tested [10, 19]. The vector $y = (y_1, y_2, \cdots, y_n)$ indicates instead the measured dataset with n being the sample size that is the number of data points. With this notation in mind, we can express the Bayes theorem [20] as follows:

$$P(\Theta|y) = \frac{P(y|\Theta)P(\Theta)}{P(y)}, \tag{2}$$

where $P(\Theta|y)$ is the posterior distribution of the parameters built according to the experimental outcome, $P(\Theta)$ is the prior distribution (or simply prior) of the parameters, that is, that in one's hands before any data measurement, $P(y|\Theta)$ is the likelihood of the data, that is, the probability of observing the data conditional on a certain parameter vector, and $P(y)$ is the marginal probability of the data, which plays the role of normalizing constant, so that Eq. (2) has a unit integral over the variable Θ. We stress that the prior probability includes all our initial knowledge (or ignorance) and can be more or less informative depending on the preliminary insight we got on the problem at hand. Bayes' theorem is therefore a prescription on how to learn from experience, insofar as it gives a golden rule to update one's beliefs in light of the accrued data.

Now let us imagine to have achieved a portrayal of $S(Q,E)$ of a given sample from a measurement run of a certain duration t_1. We can ideally try to fit this first rough $S(Q,E)$. This will provide a first joint multi-dimensional posterior distribution of the parameter vector Θ, which likewise improves our knowledge of model parameters with respect to the prior we started with. It is meaningful to think about this posterior as an updated prior to feed back into the Bayes theorem, as we keep gathering new data in the experiment.

Unfortunately, the posterior distribution has no explicit analytical expression, thus being of hardly any use to feed back in the Bayes theorem again. We then measure a second run, which, with no loss of generality, can be assumed for simplicity of the same duration $t_2 = t_1$ of the first one. New data can be certainly add to the old ones to get a new, more accurate, dataset.

Upon indicating data gathered during the first and second run, respectively, as $y = (y_1, y_2, \cdots, y_n)$ and $y' = (y'_1, y'_2, \cdots, y'_n)$, we can formally express the posterior distribution of the parameters vector, conditionally on the complete collection of data, as

$$P(\Theta|y, y') = \frac{P(y'|\Theta, y)P(\Theta|y)}{P(y'|y)} \tag{3}$$

which is a mere formulation of the Bayes theorem. We observe how the prior we have now is just the posterior distribution for the vector parameter Θ having already observed the dataset y.

The datasets y' and y being independent, we have that $P(y'|\Theta, y) = P(y'|\Theta)$ and $P(y'|y) = P(y')$. On the other hand, we can apply again the Bayes theorem to get $P(\Theta|y) = P(y|\Theta)P(\Theta)/P(y)$. Doing the substitutions, Eq. 3 becomes:

$$P(\Theta|y, y') = \frac{P(y'|\Theta)P(y|\Theta)P(\Theta)}{P(y')P(y)} \tag{4}$$

Once again, y' and y being independent, we have $P(y'|\Theta)P(y|\Theta) = P(y', y|\Theta)$ and $P(y')P(y) = P(y', y)$ for the property of the joint probability of independent variables. Eq. (4) becomes:

$$P(\Theta|y, y') = \frac{P(y, y'|\Theta)P(\Theta)}{P(y, y')} \tag{5}$$

We finally observe that the posterior probability for the vector Θ given the datasets y' and y can be obtained *via* Eq. (3) provided the posterior we derived after the first measurement is used as a new prior. This is equivalent to using as a prior the one we started with, yet multiplied for the likelihood pertinent to (inclusive of) all data collected thus far.

We can thus apply Eq. (5) in a recursive fashion to analyze on the fly neutron scattering data as we collect them. We would like to determine the most appropriate total acquisition time based on solid statistical arguments and with the prospect of inferring something about the quality of the data collected. The ultimate goal would be to have the possibility of ending the acquisition when the maximum level of information that can be obtained from the measurement has already been reached. Further prolonging the acquisition would not bring any extra relevant information. Certainly deciding when the measurement can reasonably be interrupted is at the discretion of the experimenter who may still want a specific precision from the measurement.

3.3 Simulation of a Brillouin neutron scattering experiment

Let us imagine to perform a BNS measurement. The instrument will acquire scattering intensity for a certain time, splitting such an acquisition in separate runs of the same duration. We can focus now on the spectrum corresponding to a constant Q cut of the $S(Q, E)$ we are measuring. To visualize this, we can generate simulated experimental data as they were actually measured. **Table 1** provides the parameters

Parameter	Value
A_e	33
Ω_1	4
Ω_2	8
Γ_1	0.8
Γ_2	1
A_1	10
A_2	5

Table 1.
Absolutes parameter values for the model from which the simulated datasets were drawn.

setting used in the simulation study. These otherwise arbitrary parameter values are chosen to reproduce a typical spectrum collected in a neutron or X-ray scattering measurement on an amorphous system. More specifically, we have tuned the parameters so to obtain barely resolved inelastic contributions to the spectrum, which is a typical problem faced in routine measurements. Indeed, because of the limited instrument resolution and finite counting statistics, whenever the excitation features are not sufficiently sharp, that is, adequately separated and broadened by small damping, they can blend to each other and partially merge into the dominant central peak, which makes their detection challenging. In these cases, the limited acquisition time of a single measurement become an even more constraining factor. Therefore, the chosen parameter set is suitable to mimic a typical scenario encountered in scattering studies on disordered systems. In **Figure 2**, we show one of these datasets randomly generated using the following simple model:

$$\tilde{S}(Q,E) = R(E) \otimes S(Q,E) \tag{6}$$

where $\otimes$ indicates the convolution product, $R(E) = \frac{1}{\sqrt{2\pi}\zeta} \exp\left(-\frac{E^2}{2\zeta^2}\right)$ is the instrument resolution function, assumed to have a zero-centered Gaussian profile with ζ^2 variance, which in the present case gives a $FWHM$ = 3.1 meV. Once again, we suppose to use the BRISP spectrometer with an incident energy E_0 = 80 meV, as achieved by using the (004) reflection from a Pyrolytic Graphite monochromator [15]. The dynamic structure factor is here approximated as:

$$S(Q,E) = A_e(Q)\delta(E) + [n(E)+1]\frac{E}{k_BT}\left\{\sum_{k=1}^{2}\frac{2}{\pi}A_k(Q)DHO_k(Q,E)\right\} \tag{7}$$

where $\delta(E)$ is the Dirac Delta function describing the elastic response of the system modulated by an intensity factor $A_e(Q)$, DHO_k are k inelastic contributions to the spectrum described by Damped Harmonic Oscillator (DHO) functions [21],

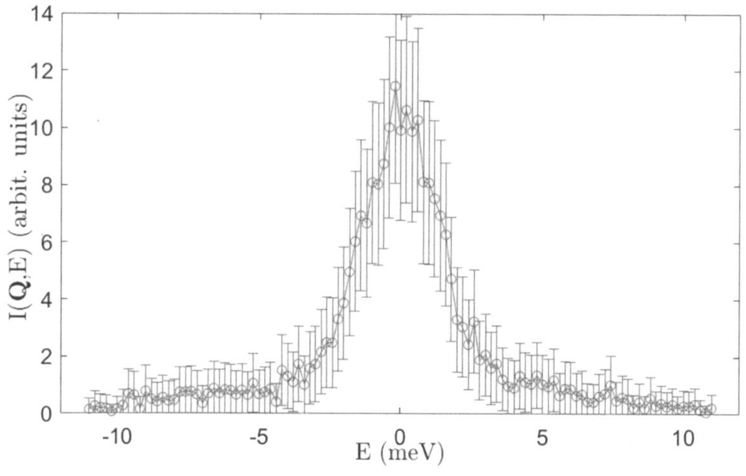

Figure 2.
Generated spectrum as drawn from the model in Eqs. (6) and (7) at a Q value of 5 nm⁻¹. This spectrum simulates the data as they could appear after a very short acquisition run. The plotted quantity is in fact the scattered intensity, to which the dynamic structure factor is proportional.

$n(E) = \left(e^{E/k_BT} - 1\right)^{-1}$ is the Bose factor expressing the detailed balance condition, k_B is the Boltzmann constant, and we have chosen the temperature T=1337 K. Finally, the simulated experimental data points are corrupted by an additive random fluctuation $\varepsilon(Q, E)$:

$$y(Q,E) = \tilde{S}(Q,E) + \varepsilon(Q,E), \quad (8)$$

with $\varepsilon(Q,E) \sim \mathcal{N}\left(0, \sigma^2\tilde{S}(Q,E)\right)$, for any Q and E, where the symblol $\sim$ means "distributed according to," $\mathcal{N}$ denotes the Gaussian distribution and σ^2 is a constant factor to be estimated on the experimental data.

We can then generate as many experimental runs as we wish and sum them as usually done. When a scattering measurement is actually performed in the large-scale facilities above mentioned, the beam time granted to researchers is probably the most critical requisite for a successful experiment. The number of equal duration measurement runs on a given sample provides the total time allotted for that sample. In **Figure 3**, we show data as they result from the sum of 20 runs of identical integration time to qualitatively visualize the improvement in data precision and spectral shape definition that can be achieved by enhancing the counting statistics through a factor 20 increase of the acquisition time.

We will try now to fit our experimental data using a Bayesian Markov Chain Monte Carlo (MCMC) [22] algorithm equipped with a Reversible Jump option (RJ) [23], as explained in detail in Refs. [10, 11]. This algorithm allows to draw values from a distribution which is only known up to a normalization constant and thus to simulate the joint posterior distribution of the parameter vector of the model, Θ, as defined in Eq. 7. The analytical evaluation of the normalization constant is in fact usually really hard if not impossible at all. We again stress that in this simplified model the number k of inelastic components contributing to the spectrum is in itself a free model parameter to be estimated conditionally on available data. Notice that the RJ option allows the MCMC algorithm to explore models with different numbers k of inelastic components with $k = 1 \ldots k_{max}$, k_{max} being the maximum number of excitations allowed. As a first step, the first measurement run is best fitted by the model and the first-level

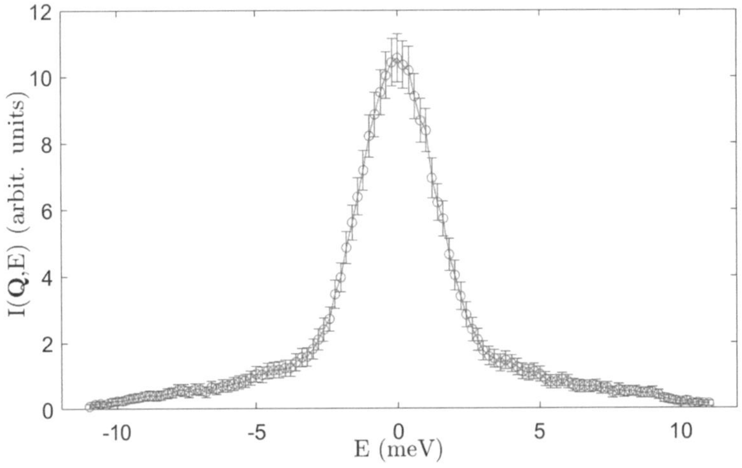

Figure 3.
Sum of 20 generated spectra as drawn from the model in Eqs. (6) and (7).

posterior distribution of the model parameters vector is thus obtained. Once the corresponding marginal distribution $P(k|y)$ has been computed, the best-fit model conditional to the first experiment run can be determined. After a second run of data y' is available, this is added to the previous one and the MCMC algorithm is used on the complete data to obtain a new joint posterior $P(\Theta|y, y')$. Parameter estimates from the first run can be used as starting values to speed up convergence. The process repeats itself as new runs become available.

Here below (**Figures 4–6**) we show the posterior distribution for some of the model parameters when the data are analyzed considering 5, 10, 20, 40 and finally 60 runs. We emphasize again that at each step we are applying the Bayes theorem, feeding back the posterior we obtained at a previous step as a prior (new knowledge about the data) for the following step. The likelihood is enriching itself more and more as long as we acquire new data. In the present example, the algorithm finds as best model the one with two inelastic modes as it should be desirable since the generation model is the one in Eq. (7). In fact because of the random error added to simulate a *real* experimental dataset, especially if the inelastic modes are chosen to have frequencies close to each other and/or large damping, it is not straightforward to find the number of modes predicted by the model whatever is the amount of data considered.

Figure 4 clearly demonstrates that the posterior distribution for the inelastic shifts Ω_1 and Ω_2 of the spectrum get sharper upon increasing the number of runs considered in the analysis. The distribution becomes better shaped, peaked, and symmetric providing, of course, a better estimate of the model parameters. Still, if we take the mean of the posterior distribution for Ω_1 (or for Ω_2) after 40 runs and we compare it with the one obtained after 60 runs, the difference between these two means is about 1%; when comparing instead the mean after 5 runs with the mean after 60 runs, the difference amounts to less than 3% which very likely is already smaller than experimental uncertainties typically reported in dispersion curves displayed by scientific papers. Similar considerations hold for the other two parameters defining the damped harmonic oscillators, namely the peak amplitudes A_1 and A_2 and the dampings Γ_1 and Γ_2 (**Figures 5** and **6**) [24].

In **Figures 7** and **8**, the best fit after 5 runs and 60 runs sum spectra is shown along with the estimated DHO components and the central elastic contribution.

In **Figure 9**, we report the posterior distribution function for the number of the detected inelastic modes as a function of the number of runs considered in the

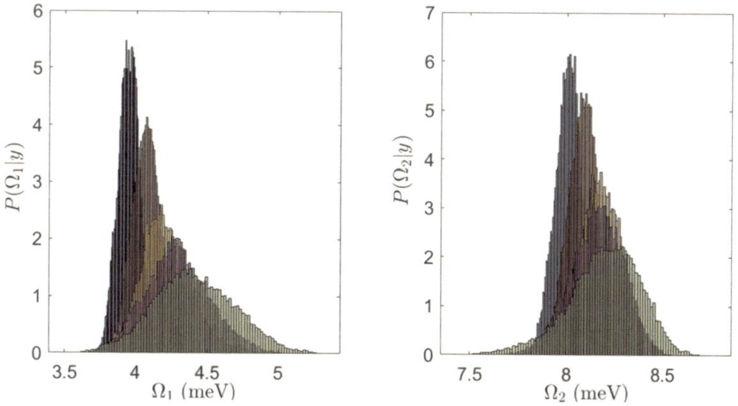

Figure 4.
Posterior distribution of the two excitation frequencies as estimated from the Bayesian analysis after 5 (green), 10 (purple), 20 (yellow), 40 (brown), and 60 (blue) experimental runs.

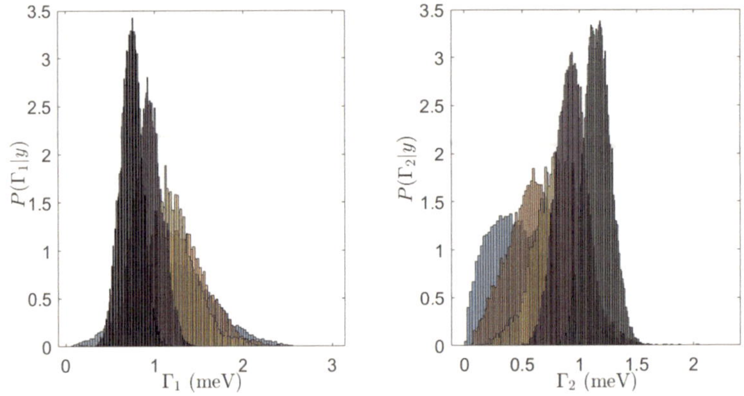

Figure 5.
Posterior distribution of the dampings of the two excitations as estimated from the Bayesian analysis after 5, 10, 20, 40, and 60 experimental runs.

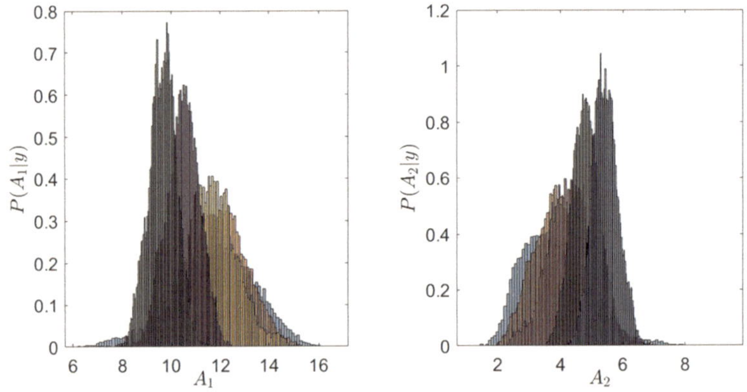

Figure 6.
Posterior distribution of the amplitudes of the two excitations estimated from the Bayesian analysis after 5, 10, 20, 40, and 60 experimental runs.

analysis. It is evident that as the experimental evidence becomes more precise, the probabilities associated with a higher number of modes progressively vanish.

To conclude we briefly draw the reader's attention on the necessity to assess convergence of the MCMC algorithm before using its output for inferential purposes. In literature, a great deal of effort has been spent in developing convergence diagnostic tools for MCMC. Some of these tools are specifically intended to check convergence of the Markov chain to the stationary distribution, or to check for convergence of summary statistics, such as sample means, to the corresponding theoretical quantities. For a recent review of the subject, see Ref. [25]. Although many convergence criteria and stopping rules with sound theoretical foundation have been proposed, in practice MCMC users often decide convergence by applying empirical diagnostic tools, in particular graphical methods. The most common graphical convergence diagnostic method is the trace plot, which is a time series plot showing the values of the model parameters at each sweep against the sweep numbers. The trace plot enables to visualize the capability of the Markov chain in exploring the parameter space. For example, the presence of flat bits reveals that the MCMC algorithm gets stuck in some part of the parameter space and is a symptom of slow convergence. This

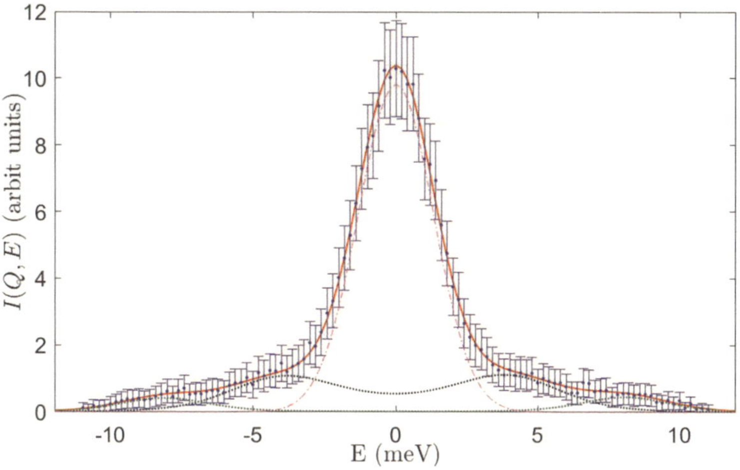

Figure 7.
Simulated spectra (blue dots) at a Q value of 5 nm^{-1} as obtained summing 5 runs drawn by the model in Eq. (7). The best-fit model (red curve) to the drawn data and the estimated DHO components (black and green line) are also shown. The dash dot magenta line is the estimated elastic central component.

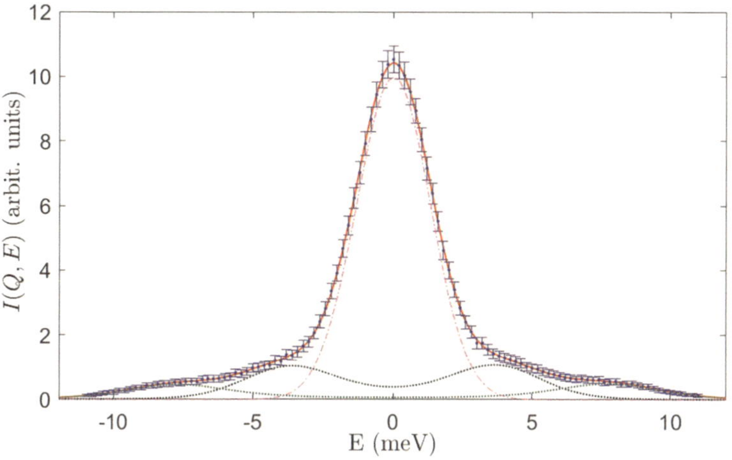

Figure 8.
Simulated spectra (blue dots) at a Q value of 5 nm^{-1} as obtained summing 60 runs drawn by the model in Eq. (7). The best-fit model (red curve) to the drawn data and the estimated DHO components (black and green line) are also shown. The dash dot magenta line is the estimated elastic central component.

happens when too many proposals are rejected consecutively. On the other hand, when proposals are too easily accepted, the algorithm may move slowly not exploring the parameter space in an efficient way. In this case, the trace plots would show visible trends or changes in spread, implying that stationarity has not been reached yet. Often Bayesian statisticians refer to a "hairy caterpillar" when describing trace plots and what they should look like. In **Figure 10**, we report trace plots for the excitation frequencies and for the number of inelastic modes in the spectrum. Another helpful graphical method is the running mean plot, which shows parameters' time-average estimates against the iterations. This line plot should stabilize to a fixed value as iteration increases (**Figure 11**).

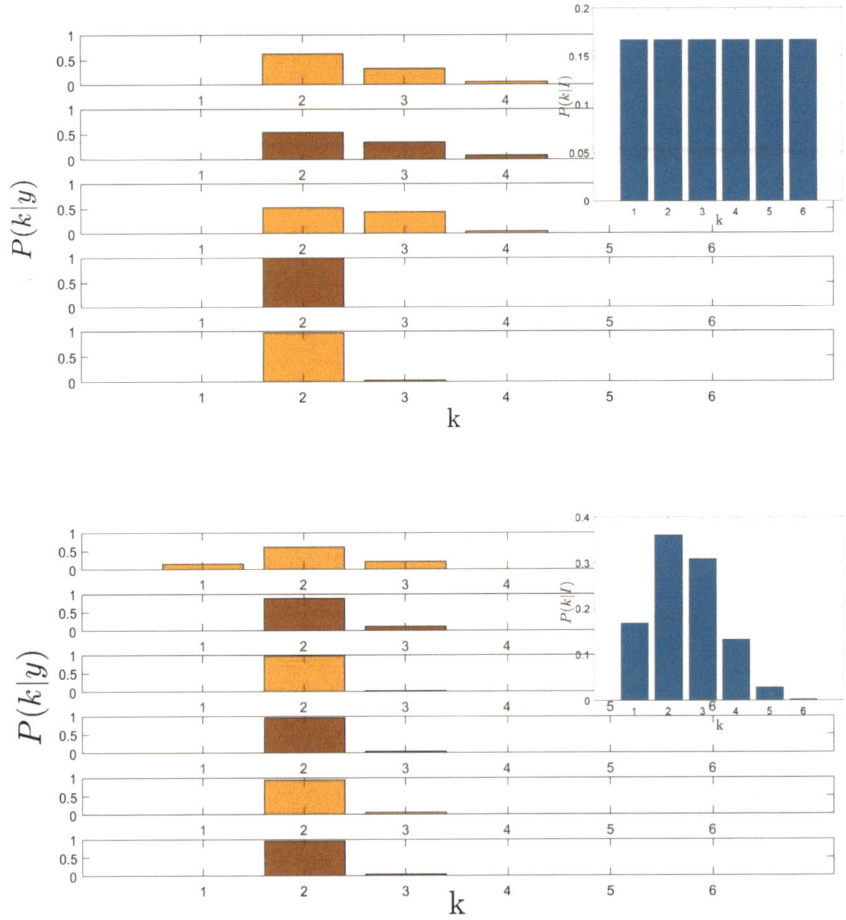

Figure 9.
Top panel: Posterior distribution function for the number k of inelastic modes detected in the simulated Brillouin spectrum as a function of acquisition time. From top to bottom, the results after 5, 10, 20, 40, and 60 experimental runs. Bottom panel: As in the top panel but at the very top of the figure the posterior of k after only 1 run also is shown. In the insets, two different priors P(k) are shown. In the top panel a uniform prior for k is plotted. In the bottom panel, a modified (see text) binomial prior distribution has been chosen for comparison.

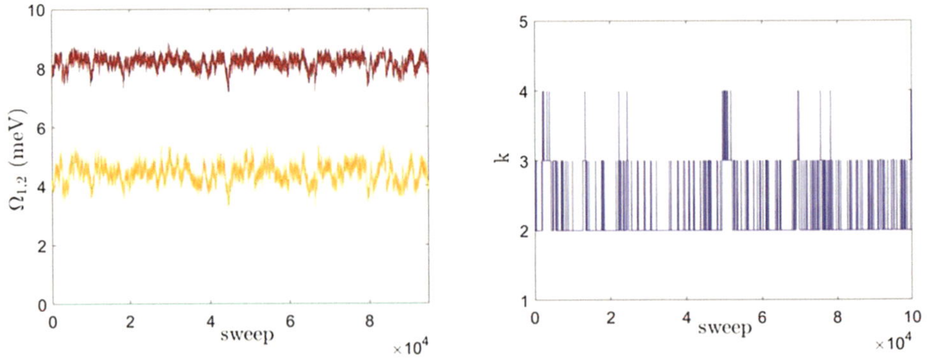

Figure 10.
Left panel: Trace plot for the excitation frequencies in the spectrum obtained summing 20 experimental runs. Right panel: Trace plot for the number k of DHOs for the same data.

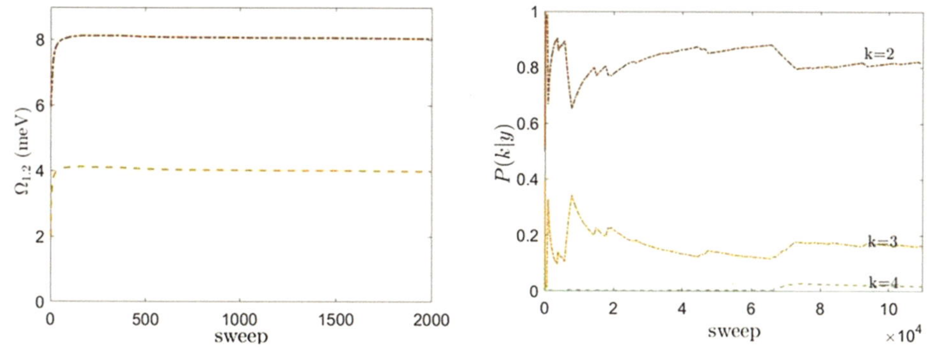

Figure 11.
Left panel: $\Omega_{1,2}$ time-average estimates as a function of algorithm sweep. Right panel: Cumulative occupancy fraction for the most visited models.

4. A few caveats and additional remarks

For some neutron scattering techniques, the raw data are not immediately available for a reliable lineshape analysis since collected intensities are affected by many spurious contributions. Indeed, the intensity ultimately detected contains, beside the genuine signal from sample, the unwanted ones coming from the empty cell, multiple scattering, sample environment, background; furthermore, it partly results from misguiding effects such as sample auto-shielding and detector efficiency. The importance or even the presence of such spurious effects changes for different neutron techniques; in this perspective, dedicated considerations and suitable adjustments to the recursive method proposed might be required. Data in the test discussed here are assumed to be already corrected for all these effects, that is, already cleaned from any unwanted contribution. Therefore, this is an ideal case, which might be not always straightforward applicable. Differently a reducing data routine has to be performed in advance. Nevertheless, depending on the technique an effort might be done to recognize some quality parameter to draw out the same conclusions that we can get from the model parameters we have seen here above. It is also true that in other neutron techniques, the possibility to reduce the data rapidly letting them available for an *on the run* analysis is preventing the problems here aforesaid and this is even more true for IXS.

Overall, the results of the test discussed are not surprising, as an improvement of the statistical accuracy expectedly enhances the precision of the parameters' determination. Nonetheless, this simple analysis shows how informed decisions about ending or continuing a measurement can be taken based on quantitative grounds. The knowledge of an entire multi-dimensional joint posterior distribution, the evolution of its mode, and overall shape upon increase of the acquisition time could help us to establish not only if data collected are sufficiently integrated but if further counting can enhance the measurement's insight. To this purpose, we can illustrate briefly an example slightly different from the one proposed before. Let us assume that the dynamic structure factor features two pairs of inelastic peaks (shoulders), for example, not only the one observed in the previous example, but, as normally the case for liquids [16, 26], an additional one which, for some reason, does not emerge clearly from the spectral shape. For instance, at low Q's, the first pair can be completely submerged by the resolution tails, while at larger Q's, the paired modes can become highly damped barely standing out from the background, while, at even larger Q's, they can move out of the energy window covered. Furthermore, if the counting statistics is poor, the marginal posterior distribution for the number of modes can convey ambiguous information and, in some instances, the presence of the second pair of inelastic modes can be overlooked. With an on-line analysis of spectra under collection, one can likely appreciate the possible evolution of such a distribution upon increasing the integration time. For instance, at the early stages of the experiment such a distribution may lead to infer a single pair of inelastic modes, while two (or more) pairs can be inferred as the measurement progresses. However, the incorporation of the Occam's razor principle [11, 27] in the Bayes theorem represents a safe antidote against the risk of overparametrization, especially when the counting statistics is still poor. This can be sufficient to keep the value of k from exceeding its true value, which in this example is known to be 2. **Figure 9** (top panel) illustrates how the posterior distribution of the number of modes k evolves as a function of the integration time. This trend has a straightforward explanation if one considers the gradually improving of statistical accuracy. At the beginning of the measurement, the algorithm could struggle to establish the true value of k assigning not negligible probability to models with a redundant number of modes. As the data are further *harvested*, the posterior distribution becomes more accurate and the number of modes converges to the most plausible one (i.e., $k = 2$). In **Figure 9** (bottom panel), we show the evolution of the posterior distribution of k as a function of the measurement acquisition time when a different prior for k is chosen. Since we have simulated experimental runs from a model with two DHOs, in this specific case, we know that the best model to fit the data must have two inelastic modes. Therefore, the chosen prior is a modified binomial distribution which privileges a solution with two inelastic modes. In this case, the prior was:

$$P(k) = \binom{k_{max} - 1}{k - 1} \pi^{k-1} (1 - \pi)^{k_{max} - (k-1)} \tag{9}$$

where k_{max} is the maximum number of modes contemplated by the model and we set $\pi = 0.3$. With this π value, the variable $(k - 1) \sim \mathcal{Bin}(k_{max} - 1; \pi)$ and the different values of the prior are reported in **Figure 9** (inset of bottom panel). In this figure, we have included also the results obtained by considering only a single experimental run.

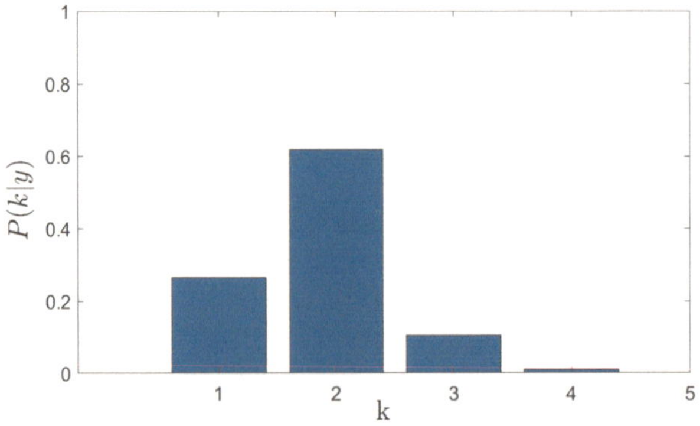

Figure 12.
Posterior distribution function for the number k of inelastic modes detected in the spectrum after only one experimental run.

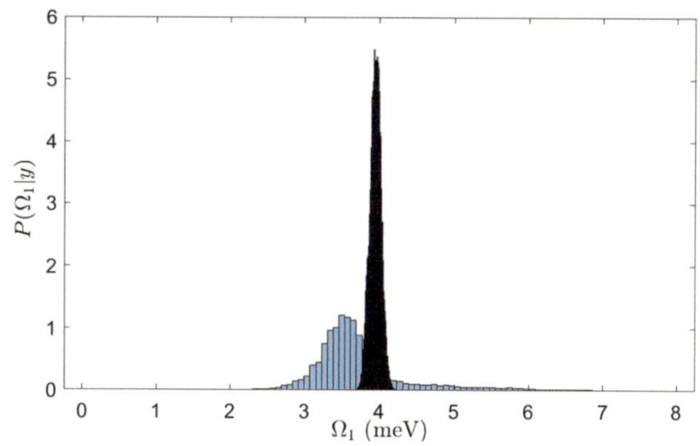

Figure 13.
Posterior distribution for the lower excitation frequency after one experimental run and 60 runs.

It appears that, when we have a firm prior knowledge about the system at hand, the posterior distribution converges to its asymptotic value even faster. In fact after five experimental runs we obtain a probability $P(k = 2)$ already close to 90%. In **Figure 12**, we show instead the values of the posterior distribution for k attained after a single experimental run. When, as in this case, the counting statistics is really poor the probabilities associated to values of k different from the expected value, that is, the one of the generating model ($k = 2$) is not negligible. Finally, the evolution of the posterior for the low-frequency excitation shift strikingly emerges from **Figure 13** in which we compare the results obtained either after a single run or after 60 runs.

5. Conclusions and perspectives

This chapter deals with a topic of pivotal interest for scattering experiments at large-scale research infrastructures, as the optimal use of the usually short beamtime allocated for the measurement. The analysis of simulated measurements presented here demonstrates that the assistance of a Bayesian inference protocol can provide a decisive advantage in the decision making and time optimization processes of routine inelastic scattering experiments, and, more in general, of any scattering or diffraction measurement. Specifically, we considered a prototypical neutron scattering study split into shorter acquisition runs; Bayesian inference is used to analyze partial acquisitions obtained by summing an increasing number of individual runs to ultimately guide the investigator in his/her difficult decision on when to stop the beam counting. Such a decision is based upon previously established success criteria, as the achieved evidence for a physical phenomenon affecting the spectral shape, or the met targets in the experimental uncertainties associated with a given lineshape modeling. In this perspective, the development of a dedicated Measurement Integration Time Optimizer protocol could be especially beneficial, as it would provide conventional neutron or X-ray investigations with real-time Bayesian inference assistance. We believe that the availability of a similar on-the-fly data analysis tool can drastically minimize the time wasted in beamtime measurement, also holding the potential for a drastic revision of the beamtime allocation process. In fact, with this novel data analysis tool, decisions on beamtime assignment can be taken on the ground of spectral simulations in which the spectra to be successively measured can be analyzed as obtained with different integration times. We anticipate that these novel inference tool can mark a discontinuity in the workflow of typical scattering experiments at large-scale research facilities.

Appendix A

Time-of-Flight neutrons instruments are a class of spectrometers which allows measuring inelastic neutron scattering, providing insights into the dynamics of matter. In **Figure 14**, we show, as an example, BRISP, a direct geometry Brillouin spectrometer once installed in the reactor hall of the High Flux Reactor of ILL. The neutrons scattered by the sample are collected by a highly pixeled detector covering a certain angular range. The scattering angle 2Θ is defined as the angle between the direct beam axis and the direction of the scattered neutrons (**Figure 15**). A Fermi chopper device splits the continuous beam coming from the monochromator into 10 μs bursts of neutrons and fixes for each burst an initial reference time. The wavevector $\mathbf{k}_0$, hence the energy E_0 of the neutrons impinging on the sample, are known and so is the time t_0 such incident neutrons take to fly from the reference initial time to the sample position. The detector electronics suitably synchronized with the chopper provides us a measure of the total time-of-flight t_{tof} from the reference time to a

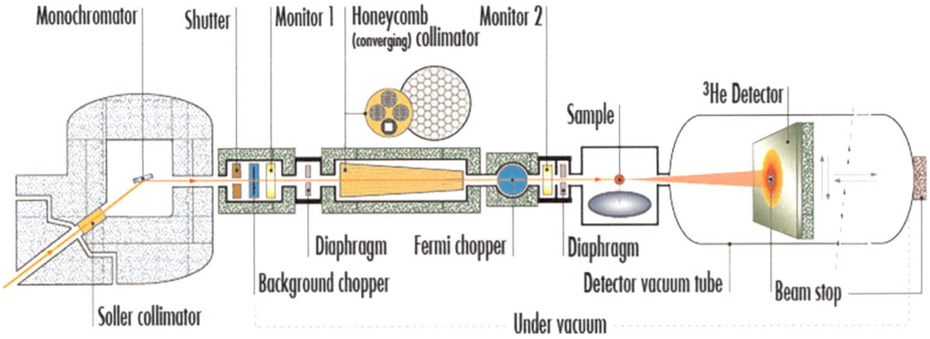

Figure 14.
Sketch of the Brillouin spectrometer BRISP. A monochromatized continuous beam, severely collimated is converted in a pulsed beam by a Fermi chopper which labels each pulse with an initial reference time as the starter in a running race. Once the neutrons interact with the sample they are scattered and finally collected by 2D-pixeled detector. Reproduced from ref. [28].

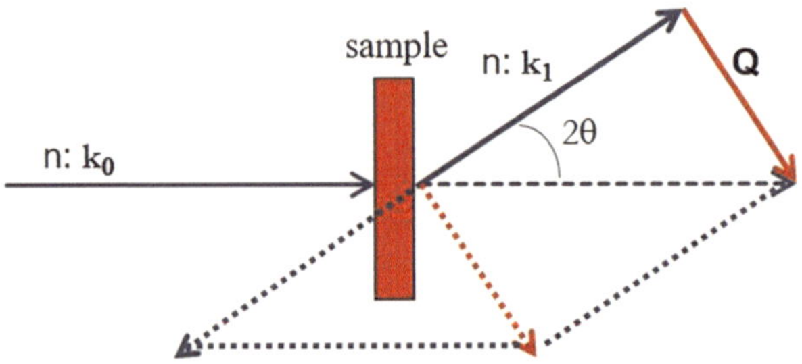

Figure 15.
Sketch of the kinematic scattering triangle. Incident neutrons characterized by a wavevector $\mathbf{k}_0$ *are scattered by a sample with a wavevector* $\mathbf{k}_1$*. The angle 2Θ between the incident and the scattered radiation is the scattering angle. The vector* $\mathbf{Q}=\mathbf{k}_0-\mathbf{k}_1$ *is the transferred momentum in the scattering process.*

specific pixel and of course the position and hence the distance traveled by each scattered neutron. If we call t_1 the time the scattered neutron takes to fly from an interacting atom in the sample to a specific detector pixel we have that:

$$t_{tof} = t_0 + t_1 = \frac{L_0}{v_0} + \frac{L_1}{v_1} \tag{10}$$

where L_0, L_1 are the distances between the chopper and the sample and between the sample and the detector, respectively, and v_0 and v_1 are the initial and final velocities of the incident and scattered neutron.

From Eq. (10) it is straightforward to obtain the energy E_1 with which the neutron reaches the detector and then the energy $E = \hbar\omega$ transferred from the probe particle to the sample. It is, in fact:

$$E_1 = \frac{1}{2}mv_1^2 = \frac{1}{2}m\left(\frac{L_1}{t_{tof} - \frac{L_0}{t_0}}\right)^2 = \frac{1}{2}m\left(\frac{L_1}{t_{tof} - L_0\sqrt{\frac{m}{2E_0}}}\right)^2 \tag{11}$$

and

$$E = E_1 - E_0 = \hbar\omega \tag{12}$$

Conflict of interest

The authors declare no conflict of interest.

Thanks

We would like to thank U. Bafile, E. Guarini, F. Formisano, R. Magli, and M. Maccarini for the very stimulating discussions. The open access fee was covered by Institut Laue Langevin (ILL), Grenoble France.

Abbreviations

BNS	Brillouin neutron scattering
MITO	Measurement Integration Time Optimizer
INS	inelastic neutron scattering
IXS	inelastic X-ray scattering
BRISP	BRillouin spectrometer
DHO	damped harmonic oscillator
MCMC	Markov Chain Monte Carlo
RJ	reversible jump
TOF	Time Of Flight

Author details

Alessio De Francesco[1,2]*, Luisa Scaccia[3], Martin Bohem[2] and Alessandro Cunsolo[4]

1 Consiglio Nazionale delle Ricerche, Istituto Officina dei Materiali c/o OGG, Grenoble, France

2 Institut Laue Langevin, Grenoble, France

3 Dipartimento di Economia e Diritto, Università di Macerata, Macerata, Italy

4 Department of Physics, University of Wisconsin, Madison, USA

*Address all correspondence to: defrance@ill.fr

References

[1] Institut Laue Langevin (ILL) [Online]. Available from: https://www.ill.eu/

[2] European Synchrotron Radiation Facility (ESRF) [Online]. Available from: https://www.esrf.fr/

[3] ISIS Neutron and Muon Source (ISIS) [Online]. Available from: https://www.isis.stfc.ac.uk/Pages/home.aspx

[4] Diamond Light Source (Diamond) [Online]. Available from: https://www.diamond.ac.uk/Home.html

[5] Swiss Spallation Neutron Source (SINQ) [Online]. Available from: https://www.psi.ch/en/sinq

[6] Swiss Light Source (SLS) [Online]. Available from: https://www.psi.ch/en/sls

[7] European Spallation Source (ESS) [Online]. Available from: https://europeanspallationsource.se/

[8] MAX IV Light Source (MAX IV) [Online]. Available from: https://www.maxiv.lu.se/

[9] Sivia D, Skilling J. Data Analysis: A Bayesian Tutorial. New York, United States: Oxford University Press; 2006

[10] De Francesco A, Guarini E, Bafile U, Formisano F, Scaccia L. Bayesian approach to the analysis of neutron Brillouin scattering data on liquid metals. Physical Review E. 2016;**94**:023305

[11] De Francesco A, Cunsolo A, Scaccia L. Chapter 2: Bayesian approach for X-ray and neutron scattering spectroscopy. In: Cunsolo A, Franco MKKD, Yokaichiya F, editors. Inelastic X-Ray Scattering and X-Ray Powder Diffraction Applications. London, UK: IntechOpen; 2020. p. 26

[12] Lovesey SW. Theory of neutron scattering from condensed matter. In: Nuclear Scattering. Vol. 1. New York, United States: Oxford University Press; 1984

[13] Windsor CG. Pulsed Neutron Scattering. London, UK: Taylor and Francis; 1981

[14] Squires GL. Thermal Neutron Scattering. Cambridge, UK: Cambridge University Press; 1978

[15] Aisa D, Babucci E, Barocchi F, Cunsolo A, D'Anca F, De Francesco A, et al. The development of the BRISP spectrometer at the Institut Laue-Langevin. Nuclear Instruments and Methods in Physics Research Section A. 2005;**544**:620

[16] De Francesco A, Bafile U, Cunsolo A, Scaccia L, Guarini E. Searching for a second excitation in the inelastic neutron scattering spectrum of a liquid metal: A Bayesian analysis. Scientific Reports. 2021;**11**:13974

[17] Guarini E, De Francesco A, Bafile U, Laloni A, del Rio BG, Gonzalez DJ, et al. Neutron Brillouin scattering and ab initio simulation study of the collective dynamics of liquid silver. Physical Review B. 2020;**102**:054210

[18] Cunsolo A. The THz Dynamics of Liquids Probed by Inelastic X-Ray Scattering. Singapore: World Scientific; 2021

[19] De Francesco A, Scaccia L, Lennox RB, Guarini E, Bafile U, Falus P, et al. Model-free description of

polymer-coated gold nanoparticle dynamics in aqueous solutions obtained by Bayesian analysis of neutron spin echo data. Physical Review E. 2019;**99**: 052504

[20] Bayes T, Price R. An essay towards solving a problem in the doctrine of chances. Philosophical Transactions of the Royal Society. 1763; **53**:370

[21] Bafile U, Guarini E, Barocchi F. Collective acoustic modes as renormalized damped oscillators: Unified description of neutron and X-ray scattering data from classical fluids. Physical Review E. 2006;**73**:061203

[22] Gilks WR, Richardson S, Spiegelhalter DJ. Markov Chain Monte Carlo in Practice. London, UK: Chapman & Hall/CRC; 1996

[23] Green PJ. Reversible jump Markov chain Monte Carlo computation and Bayesian model determination. Biometrika. 1995;**82**:711

[24] De Francesco A, Scaccia L, Maccarini M, Formisano F, Zhang Y, Gang O, et al. Damping off terahertz sound modes of a liquid upon immersion of nanoparticles. ACS Nano. 2018;**12**: 8867

[25] Vivekananda R. Convergence diagnostics for Markov chain Monte Carlo. Annual Review of Statistics and Its Application. 2020;**7**:387

[26] Cunsolo A, Suvorov A, Cai YQ. The onset of shear modes in the high frequency spectrum of simple disordered systems: Current knowledge and perspectives. Philosophical Magazine. 2016;**96**:732

[27] MacKay D. Information Theory, Inference and Learning Algorithms. Cambridge, UK: Cambridge University Press; 2003

[28] Aisa D, Aisa S, Babucci E, Barocchi F, Cunsolo A, De Francesco A, et al. BRISP: A new thermal-neutron spectrometer for small-angle studies of disordered matter. Journal of Non-Crystalline Solids. 2006;**352**:5130